SPACES OF MASCULINITIES

Changing circumstances in Western and global societies have introduced new constraints and opportunities for men and the formation of male identities. Meanwhile, the emerging diversity of masculine identities, which depart from traditional conceptions of middle-class, white, heterosexual men, poses new challenges for the production and use of spaces. *Spaces of Masculinities* provides a comprehensive introduction to the innovative and diverse research on spaces of masculinity. Drawing on a variety of geographical research projects, the central concern of the book is to highlight the significance of research about masculinity in sociological and geographical work dealing with constructions of gender.

Bettina van Hoven is Lecturer in Cultural Geography at the University of Groningen, The Netherlands, and **Kathrin Hörschelmann** is Lecturer at the Department of Geography, University of Durham.

CRITICAL GEOGRAPHIES
Edited by
Tracey Skelton, *Lecturer in Geography, Loughborough University*, and
Gill Valentine, *Professor of Geography, The University of Sheffield.*

This series offers cutting-edge research organized into three themes of concepts, scale and transformations. It is aimed at upper-level undergraduates, research students and academics and will facilitate inter-disciplinary engagement between geography and other social sciences. It provides a forum for the innovative and vibrant debates which span the broad spectrum of this discipline.

SPACES OF MASCULINITIES

*Edited by Bettina van Hoven and
Kathrin Hörschelmann*

Routledge
Taylor & Francis Group

LONDON AND NEW YORK

First published 2005
by Routledge
2 Park Square, Milton Park, Abingdon, Oxon OX14 4RN

Simultaneously published in the USA and Canada
by Routledge
711 Third Avenue, New York, NY 10017

Routledge is an imprint of the Taylor & Francis Group

Transferred to Digital Printing 2005

First issued in paperback 2013

© 2005 Selection and Editorial Matter, Bettina van Hoven and Kathrin
Hörschelmann; individual chapters, the contributors

Typeset in Perpetua by Wearset Ltd, Boldon, Tyne and Wear

British Library Cataloguing in Publication Data
A catalogue record for this book is available from the British Library

Library of Congress Cataloging in Publication Data
A catalog record for this book has been requested

ISBN13: 978-0-415-30696-6 (hbk)
ISBN13: 978-0-415-85994-3 (pbk)

CONTENTS

CONTENTS

CONTENTS

ILLUSTRATIONS

Figures

Tables

CONTRIBUTORS

Stuart C. Aitken is a Professor of Geography at San Diego State University. His books include *Geographies of Young People: The Morally Contested Spaces of Identity* (Routledge, 2001); *Family Fantasies and Community Space* (Rutgers University Press, 1998); *Place, Space, Situation and Spectacle: A Geography of Film* (co-editor with Leo Zonn, Rowman and Littlefield, 1994); and *Putting Children in Their Place* (1994, Washington, DC: Association of American Geographers). He has also published widely in academic journals and edited collections on qualitative methods, film, critical social theory, children, families and communities.

David Bartol is a graduate student at the Department of Sociology and Anthropology, Carleton University, Ottawa, Canada. His research interests include resource dependency, economic development, state theory, and the labour market.

Kath Browne is a lecturer at the University of Brighton. She completed her PhD entitled 'Power, performativity and Place: Living outside of heterosexuality' in 2002. This explored non-heterosexual women's everyday lives in relation to feminist poststructural theory. She has published articles on feminist methodology and 'the toilet problem' where women are mistaken for men, and has others under review/in progress that examine: LGBTQ pride; non-heterosexual women's perceptions of urban utopias; and heterosexist discourses.

Hana Červinková teaches cultural anthropology and is Director of the International Institute for the Study of Culture and Education at the University of Lower Silesia in Wrocław, Poland. Her doctoral thesis, *We're Not Playing at Being Soldiers* (2003, New School for Social Research, New York), is an ethnography of the Czech military and of social-military relations in the post-socialist Czech Republic. Her research interests include gender, the military and identity. Recently, Hana has focused on using ethnographic research methods as a participatory tool for teaching adult students.

Paul Cloke is Professor of Human Geography at the University of Bristol. He has 20 books and over 150 papers on issues relating to geographies of the rural, and is

Founder Editor of *Journal of Rural Studies*. His recent research has focused on issues of poverty and homelessness in rural areas. He is currently producing three books (*Handbook of Rural Studies*, with Marsden and Mooney; *Country Visions*; and *Myth and Rural Culture*, with Phillips) which will add significantly to the Rural Studies literature.

Steve Derné is Associate Professor of Sociology at SUNY – Geneseo. His *Culture in Action* (1995) explores the interconnections between culture, family, gender, and emotion in North India. His *Movies, Masculinity and Modernity: An Ethnography of Men's Filmgoing in India* (2000) considers how film viewing shapes family, emotions, sexuality, and male dominance in India. In 2001, he completed research in India to understand how globalization may have changed gender culture in India. In 2002, he was a Rockefeller foundation fellow at the University of Hawaii.

Thomas Dunk is Chair of the Department of Sociology at Lakehead University. He is the author of *It's a Working Man's Town: Male Working-Class Culture in Northwestern Ontario* as well as articles on working-class culture, racism, and environmentalism. His current research focuses on masculinity, class, and the political and symbolic economy of hunting.

Rhys Evans is currently a core research fellow at the Arkelton Centre For Rural Development Research at the University of Aberdeen, in Scotland, where he is carrying out research into discursive interventions in the creation of place-meanings through the Community Woodland movement in Scotland. His research has focused upon men's relationships with feminism and, more specifically, the constitution of regimes of masculinity, especially those which grow around work practices. Key publications include chapters in edited collections on forestry and rurality and, more specifically, on masculinity *You questioning my manhood, boy? Masculine identity, work performance and performativity in a rural staples economy* (2000) (Arkleton Research Paper No. 4, Aberdeen).

C. Michael Hall is Professor of Tourism at the University of Otago. Former positions include periods at the University of Canberra and the University of New England, Australia. He is co-editor of Current Issues in Tourism and chairperson of the IGU Study Group on Tourism, Leisure and Global Change. He is the author of numerous books and articles in the fields of tourism, leisure, geography, heritage, and environmental history. Current research foci include gastronomy, economic restructuring and tourism, place promotion, second homes and various aspects of the concept of mobility.

Kathrin Hörschelmann is a lecturer in Human Geography at the University of Durham, U.K. Her doctoral research has focused on media practice and changing identities in (former) east Germany as well as on the transformation of gender relations since German unification. She is currently undertaking further research on the social and cultural consequences of post-socialist transformation and on the influence of globalization on youth cultures. Publications include papers in international journals such as *Transactions of the Institute of British Geographers*, *Political Geography*, and *Journal of Social and Cultural Geography*.

Bettina van Hoven is researcher and lecturer at the Department of Cultural Geography, University of Groningen, the Netherlands. Her research interest focuses on gender, exclusion and identities. Recent publications include the book *Made in the GDR: the changing geographies of women in the post-socialist rural society in Mecklenburg-Westpommerania* and articles in *Area*. She edited the book *Lives in transition* and currently works on the experience of place in prisons in the US, the Netherlands and eastern Germany.

Janine Janssen is a cultural anthropologist, who has specialized on research in prison. She has done research among Latin American female detainees in a Dutch prison for long-term detention. Furthermore she has written a PhD thesis on the effects of short-term imprisonment in the Netherlands. She is currently employed as a researcher for the 'Police Haaglanden' in The Hague, Netherlands. Her publications include *Latijnsamerikaanse drugskoeriersters in detentie: Ezels of zondebokken?* (Latin American drug couriers in detention) (1994) and *Laat maar zitten. Een exploratief onderzoek naar de werking van de korte vrijheidsstraf* (Just sit it out. An exploratory study into the consequences of short-term imprisonment) (2000).

Michael Johnson is the Dean of the School of African and Asian Studies at the University of Sussex. He is the author of *All Honourable Men: The Social Origins of War in Lebanon* (London: I.B. Tauris, 2001), a book that uses comparative material to analyse communal violence in Lebanon's civil wars.

Robyn Longhurst is Chairperson of the Department of Geography at the University of Waikato in Hamilton, New Zealand. She is author of *Bodies: Exploring Fluid Boundaries* (2001) and a co-author of *Pleasure Zones: Bodies, Cities, Spaces* (2001). Her research coalesces around themes of bodies, spaces and power.

Karen Lysaght is a social anthropologist. She completed her PhD research on political movements within loyalism in post-ceasefire Northern Ireland, concentrating on questions of political leadership, power, violence and territoriality. This research represents one strand of the wider 'Violence Research Programme', a British-wide collaborative study of the effects of violence on society, headed by Professor Elizabeth Stanko, Royal Holloway, University of London. The project utilizes a combination of quantitative and qualitative methodologies to examine the effects of violence and sectarianized geography on everyday life in Belfast.

Linda McDowell is Professor of Economic Geography at University College London and Director of Graduate Studies in the Department of Geography. She is the author of numerous papers about economic change and the cultures of labour, as well as feminist theory and methods. Among her books are *Capital*

Culture: Gender at Work in the City (1997) *Gender, Identity and Place* (1999) and *Redundant Masculinities?* (2003), a book about young men, masculinity and employment change. In 2005, UCL Press is publishing her book about women migrant workers in post-war Britain, provisionally entitled *Hard Labour*.

Louise Meijering is working on her doctoral thesis on 'isolated communities in rural areas' at the Department of Cultural Geography, University of Groningen, the Netherlands. Her research interest is on exclusionary geographies, rurality and identity. In a recently completed research project she studied the identities of Indian IT-professionals in Germany.

Gill Valentine and **Tracey Skelton**. Gill Valentine is a Professor of Geography at Sheffield University. Dr Tracey Skelton is a Lecturer in Geography at Loughborough University. Both have researched and written widely on questions of gender, sexuality and youth. They are co-editors of *Cool Places: Geographies of Youth Cultures* (Routledge, 1998).

Katie Willis is a senior Lecturer in Geography at Royal Holloway University of London. Her research interests focus mainly on gender, households and migration in 'The South', although she has also conducted research on migration flows of highly-skilled Britons and Singaporeans to China. She has conducted fieldwork in Mexico, China, Singapore and California. She co-edited (with Brenda Yeoh) the Edward Elgar volume *Gender and Migration* (2000) in the International Library of Studies of Migration. She is the author of *Theories and Practices of Development* (Routledge, 2003).

ACKNOWLEDGEMENTS

Several people have contributed to the making of this book during the past years. Andrew Mould is, of course, the key person in this undertaking and we would like to thank him for his help and his patience. We would like to acknowledge and thank the Tate Gallery for letting us use Millais' painting 'the North West Passage' (Chapter 1). Last but not least, we kindly thank John T. Linnell, Fionnuala Murphy and R.J. Spencer for their photos, and the authors who supplied photos and illustrations for use in their chapters.

1

INTRODUCTION

From geographies of men to geographies of women and back again?

Bettina van Hoven and Kathrin Hörschelmann

Summary

In this introductory chapter, we sketch an outline of the historical move from geographies written by (and most often about) men, to the feminist critique of gendered power relations in geographical thought and the recent emergence of research on the geographies of masculinities. We offer a summary of different theoretical approaches to the definition of masculinity and explain why, in this book, we focus on spaces of masculinities rather than men. In the final section, we give a brief overview of the structure of the book.

- Geographies of men
- Geographies of women
- Geographies of masculinities
- Defining masculinities
- Outline of 'Spaces of Masculinities'.

Geographies of men

In 1996, the Dutch geographers Ben de Pater and Herman van der Wusten provided a 'sketch' (as they called it) of the development of geographical sciences in their book *Het Geografische Huis* (the Geographical House). The two authors discuss 'the geographical roots of geographical thinking' (p. 9, our translation), the development of geography within the social and academic context and the emergence of new points of reference that stimulated several geographical specializations. From this discussion, an interesting image of geography and geographers of the early days emerges: geography has long been a discipline dominated by men and one about men. Classical, geographical thinking was, of course, highly influenced by great philosophers such as Aristotle who claimed that:

Again, the male is by nature superior, and the female inferior; and the one rules and the other is ruled; this principle of necessity extends to all mankind. Where then there is such a difference as that between soul and body, or between men and animals (as is the case of those whose business it is to use their body, and who can do nothing better), the lower sort are by nature slaves, and it is better for them as for all inferiors that they should be under the rule of the master.

(Halsall 1998)

This perception of females as inferior has remained dominant for centuries. In the early seventeenth century, for example, 'woman' was symbolic of 'Other' and seen as naturally impure (in part because she menstruates). 'Woman' was seen as sinful temptress. This perception was captured in artistic impressions and symbolism as well. Jan van der Stralt's etching 'Vespucci landing in America' (1619) illustrates that 'man' is rationality, reason and ruler. In the etching, 'man' is named (Vespucci) and empowered to name the unknown 'Other' himself. Other (America) is depicted as a naked woman, lounging in a hammock. Her nudity and leisure evoke images of sexuality and innocence, pleasure and nature. The etching is part of the art collection of the Université de Liège in Belgium and can be viewed on the university's website at www.ulg.ac.be/wittert/fr/flori/opera/vanderstraet/vanderstraet_reperta.html.

Roughly from the renaissance onwards, geographical endeavours were guided by a political need for exploration, discovery and conquest. In 1899, there were 62 organizations worldwide and several chairs in geography devoted to the subject of discovery. The work of the Royal Geographical Society (RGS) in the UK was equally marked by exploration and discovery, and later imperialism.

The relationship between British imperialism, British masculinity and British dreams of adventures into the 'geography of the unknown' can be illustrated through yet another piece of art, the painting *The North West Passage* by Millais (1874):

The North West Passage was the sea route round the North of the American continent which was thought to form a passage to China and the East. Several Arctic expeditions had tried and failed to find it. Millais painted this picture in 1874 at the height of Empire and just when another English expedition was preparing to set off on the same voyage. When the picture was shown at the Royal Academy in 1874, it was accompanied by the lines: 'It might be done, and England should do it'. This patriotic message is underlined by the presence of the Union Flag draped over a screen.

(Peart 2003)

It must be noted that the painting (see Figure 1.1) did not depict the *actual* passage but instead an old adventurer who might travel the passage in his mind. His expression is outward, the telescope and map in the sitting room and the ship on the horizon, seen through the window, are illustrative of this.

Also seen in the painting is a young woman, seated on the floor next to the old

Figure 1.1 Painting Millais (*source:* © Tate, London 2004).

man. In contrast to the adventures the man might be experiencing, she is turned inward, reading a book, without a care in the world. At the time the painting was made, women's position in science was negligible. It is interesting to note in this context that women were not permitted membership of the RGS until 1913. A significant inhibitor was the way in which women were seen to write geography. There were several women geographers who travelled and who preserved their observations in the form of diaries and travel descriptions. Isabella Bird, for example (incidentally the first female member of the RGS), published several books on her many journeys such as *An Englishwoman in America* (1856), *A Lady's Life in the Rocky Mountains* (1879) or *Unbeaten Tracks in Japan* (1880), to name but a few. However, up until the present, her contribution to geography has been questioned. A relatively recent response by Stoddart (1991) to Bird's work illustrates this:

> Simply for a person to travel about the world in the late nineteenth century did not by emerging standards of the time constitute a contribution to the discipline of geography. I am not aware, for example, that Isabella Bird ever made a measurement, a map or a collection, or indeed ever wrote other than impressionistically about the areas she visited.

(p. 484)

This quote is particularly remarkable, since by the 1990s, feminist geography had become a well-established subdiscipline and its underlying methodology accepted as a valid mode of enquiry.

Geographies of women

In spite of some existing geographical texts about and by women since the nineteenth century, it was not until Zelinsky et al. (1982) argued for a 'substantial shift in the angle of vision' as 'women . . . exist under much different conditions and constraints [than men]' (p. 353) that the neglect of 'the other half' (Monk and Hanson 1982: 11) in geography was seriously and rigorously addressed. Zelinsky et al.'s (1982) review and prospectus on the status of women in geography has widely been regarded as the 'foot in the door' after which both descriptive and theoretical discussions of women (and other previously silenced social groups) in geography were intensified and diversified.

In the 1990s, feminist thinking became strongly established in disciplines other than geography. This interest was evident in the research process itself, i.e. the feminist, political agendas of research and research methods as well as the perceived need to challenge previous perceptions of knowledge, and of who can 'know' and act. Social geographers have, for example, critiqued the gendering of urban and rural space, such as the concentration of single-parent families in poor neighbourhoods (Winchester 1999), gendered fear of crime (Pain 2001) or unequal access to places depending on gender (Peake 1993; Bondi 1998). Urban planning has been shown as a practice of materializing gendered assumptions about the use of urban space (Fincher and Jacobs 1998), while rural geographers have drawn attention to the role of women in farming and to their socio-cultural positioning in rural community life (Whatmore et al. 1994; Little and Panelli 2003). Significant attention has also been paid to the gender implications of economic change (Massey 1994; McDowell 1997) and to the gender dimensions of development (Radcliffe 1998; Momsen 1991, 1993). Engagement with queer theory has finally led some geographers to go beyond analyses of the gendering of space in order to deconstruct heterosexist assumptions and to make visible the exclusion of people with non-conformist sexualities through the construction of space (Bell and Valentine 1995; Knopp 1994; Berg 2001, 2002). Due to the increasing political involvement implied by feminist research, a central concern indeed became 'difference' and 'otherness' such as class, 'race', sexuality, or disability.

In the second half of the 1990s, feminist research was taken further by a personal commitment of the researcher to social and political change, and empowerment of the research subject. Therefore, some feminist geographers aimed to include the 'Self' in the interpretation of that part of the social world they wished to understand largely for methodological and epistemological reasons. The methodological repositioning of scholars was, initially, a controversial issue. However, this shift also contributed finally to the recognition of achievements and knowledge production of early women geographers. The debate between Domosh (1991a, 1991b) and Stoddart (1991) on the necessity of a feminist historiography of geography, for example, shed light on those women who, noticed and unnoticed, advanced the discipline through explorations or academic merits. In spite of earlier acknowledgements of contributions made by women geographers such as Isabella Bird, historic accounts by women travellers in the nineteenth or early twentieth century were discovered as components in the production of geographical knowledge only as late as the 1990s. This belated recognition of early achievements by women travellers and explorers is largely due to an (ongoing) discussion of what constitutes geographical knowledge and 'valid' contributions to the discipline as was discussed above.

In sum, a key concern for geographers in all the above work has been to show how gender identities are lived and constructed in different cultural settings. Space has been shown to be gendered in many ways, while gender itself is seen to be constructed through spatial relations and geographical imaginations (McDowell 1998; Bondi 1990; Rose 1993; Blunt and Rose 1994; Laurie *et al.* 1999; Women and Geography Study Group 1984).

Geographies of masculinities

There is no question then about the continued value and necessity of feminist geographical research today. Analysing the connections between space, place and identity remains a significant task, if we seek to transform unequal relations of gender and sexuality as they are materialized and lived spatially. But while feminist geographers have critiqued the discipline's neglect of women's experiences effectively, there has been a notable lack of attention to the formation of masculine identities and spaces. A book by geographers on spaces of masculinity is then perhaps not so surprising. In the words of McDowell (2001: 182):

> gender is not an attribute solely possessed by women. . . . It also provides an intellectual and research challenge to the one-dimensional man, garbed in his unyielding patriarchal power, by insisting that masculinity, too, is also an uncertain and provisional project, subject to change and redefinition. Geographers have perhaps, however, been slow to accept this challenge . . . relying too heavily on a singular masculinity, defined as the unchanging 'One' against which multiple and contested femininities are constructed.

A focus on the relational formation of male identities and masculine spaces seems long overdue in both feminist and gender-oriented geographical work. This has been recognized in recent work by geographers such as Peter Jackson (1994, Jackson *et al.* 2001), David Bell (2000a, 2000b; Bell and Valentine 1995), Linda McDowell (2002, 2003), Frank Mort (1996), Doreen Massey (1994), Robyn Longhurst (2000, 2001), Ruth Liepins (2000), Lise Saugeres (2002), and H. Campbell and M. Bell

(2000). Many of these authors start from the recognition that recent socio-economic and political changes have placed pressures on (traditional) male identities that confront gender research with new scope for investigation and discussion (see McDowell 2003; Massey 1994; Little 2002). Paid employment and the assumed role of men as breadwinners and guardians of their families have traditionally been key denominators for male identity. As the roles of women have changed, leaving new spaces for male identities, it has increasingly been recognized that the younger generation of men and boys cannot easily be fitted into such 'old-fashioned' categories. In addition, growing tolerance towards different sexual preferences and a decline in traditional conceptions of men as needing to be aggressive and strong (both physically and mentally) have opened up further spaces in which men may seek their identities. At the same time, however, the creation of new social norms and values has not automatically led to an abandonment of the old, thus representing significant fields of conflict for the development of identities.

One of the aims of this book is to examine the implications of these social and cultural changes for our understandings of men, masculinities and space. Before we can begin our analysis, however, we need to define that slippery concept: masculinity. What exactly is it that we are talking about? Is this a book only about men, or about maleness, masculine identities, discourses of masculinity? In order to answer these questions, we will take a look at the ways in which masculinity has been defined in the literature in the next sections.

Defining masculinities

Research on masculinities has become a prominent part of gender studies over the past 20 years. Fuelled in part by popular fears of a 'crisis' at the heart of modern masculinity, as noted above, work on gender relations has focused more carefully on the question of how masculine identities are constructed and (re)produced. It engages critically with, and develops further, approaches from psychoanalytic theory, sex role theory, feminist theory and, more recently, social constructionism and post-structuralism. In order to understand how masculinity is defined by gender theorists today, we provide a review of how the subject has been dealt with in the past, discussing each of the above approaches in turn.

Psychoanalytic and sex role theory

A common theme in psychoanalytic and sex role theory is the rather unquestioning assumption of essential differences between men and women that are reflected in masculine and feminine identities. Both theories undervalue the power relations that position men and women differently and develop a rather static, prescriptive image of gender identities. Psychoanalytic theory assumes an underlying biological reality of drives that men and women equally possess. A key proposition is that gender is acquired through an Oedipal process of identification and the repression of instinctive desires (Edley and Wetherell 1996). More recently, psychoanalytic theory has emphasized the conflicts and uncertainties that arise for the male psyche from this gendered socialization in childhood, arguing that it leads to a constant need to prove oneself as male (ibid.).

Sex role theory, similarly, constructs masculinity as the social expression of differences between men and women. In this approach, sex roles are seen as the prototypes into which men's and women's identities are forged through socialization. Men are measured by their success or failure to comply with social norms of masculinity, such as to be strong, successful, capable, reliable and in control, which Kimmel (1994) has identified as the key characteristics of 'marketplace man' (also see Brannon 1976; Seidler 1989). Sex role theory examines how a range of social institutions turn 'the boys' into 'men', e.g. the home, school, the community, peer groups, the media, the demands of the economic system. Masculinity is seen as a performance on a preset stage, leaving only a few directions to be followed (Beynon 2002). The connection between maleness and masculinity is portrayed in a one-dimensional, unidirectional way, so much so that it becomes questionable whether the distinction between maleness and masculinity makes any sense at all. Men can easily appear as victims of the process of socialization, leaving the question of responsibility and of the embodiment of power open to debate.

Feminist theory

By contrast, feminist politics and theory have been active in challenging hegemonic understandings of gender and in critiquing gendered power relations. Feminist scholars have launched a major critique of the invisibility of masculinity as a normative concept. They have problematized the assumption that research is gender neutral and shown that hegemonic masculinities are most frequently produced and reproduced in academic work as an implicit standard against which various groups of 'others' are assessed. Early feminist work sought to expose the power relations between men and women by revealing the patriarchal structure of (western, capitalist) society (Cockburn 1984; Walby 1990). While the critique of patriarchy was instrumental in bringing to light gendered inequalities arising from an uneven system of production and reproduction, this approach has tended to construct an image of masculinity that portrays it too restrictively as an instrument of oppression and dominance. Different forms of masculinity, power relations between women and queer sexualities remained undertheorized until the emergence of post-colonial, post-structural and queer theories. We will return to these more explicitly below, but for the moment, it is important to note that feminist work has been crucial for the development of critical 'men's studies' in as far as the latter seek to deconstruct homogeneous understandings of gender and sexuality and aim to challenge gendered power relations.

Critical men's studies

Authors such as Arthur Brittan (1989), Bob Connell (1995, 2000), Máirtín Mac an Ghaill (1994, 1996), Lynne Segal (1990) and Stephen Whitehead and Frank Barrett (2001a) have been at the forefront of research that seeks to understand how masculinities are constructed (rather than given a priori) and how power relations between men and women as well as among men are sustained. In accepting the feminist call for research that challenges gender inequalities, critical men's studies have resisted the temptation of bemoaning generalized threats to male identities. They critique the mythopoetic men's movement, that seeks to reassert male identities in an essentialist, homogenizing and frequently anti-feminist way (see, for example, Bly 1990). Instead, critical men's studies seek to explain more fully the relational construction of masculinity and the power relations sustained by its hegemonic definitions. Bob Connell has been quick to point out that there is no such thing as a homogenous, universal 'masculinity', but rather a wide range of masculine identities that are hierarchically structured around hegemonic understandings. He defines hegemonic masculinity as 'the configuration of gender practice which embodies the currently accepted answer to the problem of the legitimacy of patriarchy, which guarantees (or is taken to guarantee) the dominant position of men and the subordination of women' (1995: 77). Significantly, Connell here recognizes the plurality of masculinities without lapsing into the relativist position where this plurality prevents one from critiquing power relations. Different masculinities are not equal, but organized hierarchically: 'Hegemonic masculinities define successful ways of "being a man"; in so doing, they define other masculine styles as inadequate or inferior' (Cornwall and Lindisfarne 1994: 3). Thus, hegemonic definitions of masculinity receive their legitimacy from the marginalization of other forms of masculinity, such as those of different social classes, ethnicities, sexualities, ages or abilities. The latter are almost always characterized as more feminine, thus highlighting the other dynamic of hegemonic masculinity: its contrast with, and assumed superiority to femininity. In recognizing the plurality of masculinities, Connell is able to deconstruct the assumed universality of hegemonic notions of 'being a man'. Divergent claims on the meaning of masculinity reveal fissures and contradictions in the hegemonic definition, which necessitates the constant reiteration of what 'a real man' is supposed to be. The fragility and historicity of hegemonic western conceptions of masculinity has been highlighted by a number of authors from different perspectives recently. Thus, Morgan (1992, 2001) and Mosse (1996) have shown how middle class notions of masculinity developed historically in the West, while others, such as Staples (1982), Hall (1997), Phillips (1997) and Rutherford (1997) have brought attention to the racist implications of these hegemonic definitions. Queer theorists, building on the work of Michel Foucault among others, have critiqued the homophobic construction of straight, white, middle-class masculinity (Bell and Valentine 1995; Halberstam 1998), while yet others have analysed the class dimension of masculinity (Willis 1977; Massey 1994; Collinson and Hearn 1996; McDowell 2002, 2003).

Although this work has achieved a major unsettling of naturalized assumptions of what it means to 'be a man', there remain some contradictions at the heart of masculinity research that need to be examined further. Máirtín Mac an Ghaill is one of those authors who have drawn attention to the unclarity of masculinity as a field of study. He is particularly critical of the way in which research on masculinities develops 'highly complex theories . . . that fail to connect with individuals' experiences' (1996: 4). The charge of unclarity seems especially justified with a view to the persistent assumption that masculinity somehow remains the property of men. Although most sociologists today would challenge the simplicity and rather unquestioning stance of the sex role model, many continue to define masculinity in relation to men, male bodies, and their difference from women/femininity. Thus, for Buchbinder (1994) and Brittan (1989) masculinity represents various (fluctuating) aspects of men's behaviour, what men do to be acknowledged as men, while Whitehead and Barrett (2001b) characterize masculinities as 'those behaviours, languages and practices, existing in specific cultural and organisational locations, which are commonly associated with males and thus culturally defined as not feminine' (p. 15). The latter state that masculinity reflects the social and cultural expectations of male behaviour rather than biology, but fail to explain how such expectations attach to male bodies, who produces them and what to make of female masculinities. In maintaining that masculinity represents different 'ways of being a man', such work continues unwittingly to portray an essentialist view of masculinity. Jeff Hearn (1996) and John MacInnes (1998) have critiqued the continued use of the term 'masculinity' for this very reason:

> The concept may divert attention from women and gendered power relations. The use of the concepts [sic] is often imprecise. Meanings stretch from essential self to deep centre, gender identity, sex stereotype, attitudes, institutional practices and so on. What is exactly meant by masculinity is often unclear. Masculinity is often a gloss on complex social processes. The concept is sometimes attributed a causal power – for example, that masculinity is said to cause a social problem, such as violence – when masculinity is rather the result of other social processes.
>
> (Hearn 1996: 213)

Hearn calls for a more precise use of the term, for a move back from 'masculinities' to 'men' (as in 'men's practices', 'men's social relations' etc.), for exploring the multiplicity of 'discourses of masculinity' and the multiplicity of masculinities as well as for the development of concepts that more accurately reflect women's and men's differential experiences of men (p. 214). Mirroring some of these concerns, MacInnes goes further to suggest that 'we cannot either invite or compel men to reform their masculinity, because it is not something which they possess as such in the first place' (MacInnes 1998: 149). To MacInnes, masculinity is the last ideological defence of male supremacy in a world that has already conceded that men and women are equal. He argues that 'masculinity as a concept obscures the analysis of social relations between the sexes' (MacInnes 1998: 59).

The question of difference

The key issue that appears to remain unresolved in studies of masculinities is the role of the body: what is the link between masculinity and male embodiment and can there be, in the words of Judith Halberstam, a 'masculinity without men' (2002: 355)? It has been one of the major achievements of feminist and queer theory to destabilize notions of gender that assume an underlying biological bedrock of sex independent from social relations. Judith Butler's work has been particularly instrumental here. She argues that the sex/gender divide constructs heterosexuality as a norm that excludes other sexual orientations as 'unnatural' and deviant: 'This exclusionary matrix by which subjects are formed thus requires the simultaneous production of a domain of abject beings, those who are not yet 'subjects', but who form the constitutive outside to the domain of the subject' (1993: 3).

To Butler, gender is performative, 'constituting the identity it is purported to be' (1990: 25). As such, it is neither pregiven by biology nor is it simply the repetition of social norms: 'Rather than assume an originary and essential sexual identity – the essence of a person's sexuality – ideas about performativity reveal the mobility, flux and dynamism of different identities in different contexts' (Blunt and Wills 2000: 159). It is this emphasis on the dynamism of identities that we need to retain in studies of masculinities. The recognition of female masculinities and male femininities destabilizes binary differences between men and women and helps to shift some of the power inequalities between and amongst them. Halberstam argues that the insistence on masculinity as the property of men reflects a 'lack of any real investment in the project of alternative masculinities and . . . an unwillingness to think through the messy identifications that make up contemporary power relations around gender, race and class' (p. 364). Eve Kosofsky Sedgwick equally makes the point that masculinity sometimes has nothing to do with men: 'As a woman, I am a consumer of masculinities, but I am not more so than men are; and like men, I as a woman am also a producer of masculinities and a performer of them' (1995: 13). Interestingly, instead of arguing for the abolishment of the term 'masculinity', both Halberstam and Kosofsky Sedgwick retain it as a working concept, yet one that has to be separated from men. Masculinity thus becomes a much more fluid term, a cultural construction of 'manliness' that can be shared and contested by both genders and that becomes embodied in a plurality of ways.

It is out of these concerns, that we have chosen 'masculinities' as the focus of our book rather than 'men' or 'maleness'. The concept rightly evokes a complexity that needs to remain at the heart of our investigation. While 'maleness' implies an identity derived from the possession of a 'male' body, masculinity can attach to bodies, objects, places and spaces well beyond the apparent confines of biology and sex. Masculinity evokes images of maleness, yet they are by no means necessarily shared by men and can, on the other hand, be adopted by or attributed to women. The concept allows the deconstruction of hegemonic gender identities and of the assumed naturality of 'sex'. It highlights difference as well as enabling us to grasp the salience of gendered power relations. Importantly for this book, it allows us to analyse and critique the ways in which places are gendered and how this affects the identities of different subjects. For us, the concept will only lose its explanatory power if inequalities of gender and sexuality are resolved and if the terms 'man/woman', 'maleness/femaleness' are deconstructed and redefined at the same time.

Outline of *Spaces of Masculinities*

This book brings together authors from different backgrounds across the social sciences to examine the ways in which masculinities are constructed in and through space as well as discussing how particular places become gendered as masculine. A major concern for us are the implications of gendered divisions of space for social inclusion and exclusion, which were discussed above as an important topic of debate in feminist geography for some time. We have divided the book into five broad sections:

- Masculinities in transition
- Masculinities and cultural change
- Masculinities and violence
- Embodied masculinities
- Sexuality and relationships.

The first part examines how masculinities have been transformed in different parts of the world due to major social, economic and political changes. Linda McDowell addresses the connections between labour market change and masculine identities, illustrating her argument through case studies of particular occupations and young working men just entering the labour market. Thomas Dunk and David Bartol focus on working-class masculine identities in a North American industrial 'hinterland' region that changed from an area of relatively secure and well-paid employment for male, blue-collar workers to one of industrial decline stimulated by global and local environmental and economic problems. Paul Cloke considers both the relevance of research on masculinities for rural geography and how transformations in the countryside affect rural men. Hana Červinková then takes us into a different corner of the world to discuss how men in the Czech military have responded to the political changes in East-Central Europe over the past 15 years.

The theme of the following part is closely connected. Here, we look at the way in which cultural changes can destabilize as well as reconfirm traditional notions of masculinity. Bettina van Hoven and Louise Meijering demonstrate this in the case of male Indian migrant workers in Germany, while Steve Derné shows how, in the face of media globalization, Indian men reassert their masculinity by insisting on women's moral purity and allegiance to tradition. Katie Willis takes a closer look at stereotypical conceptions of Latin American men as *machos* and argues that this image hides a plurality of male subject positions. She shows that recent changes in socio-economic conditions have led to a range of different responses by Latin American men.

Part 3 engages with a difficult topic in the study of masculinities. Aggression and violence are often seen as the flip-side of stereotypical masculine identities. While social biologists would see these characteristics as originating from the male body, the authors in this book take a broadly social constructionist position and show how specific historical situations lead to the adoption of aggressive attitudes in order to assert one's masculinity, often in response to a perceived threat to the latter. Michael Johnson examines the link between ethnic violence, rural-urban migration, the transformation of family structures and the emergence of nationalism in Lebanon, while Karen Lysaght presents an analysis of the causes and consequences of young men's violent actions in Northern Ireland. In the final chapter of this part, Kathrin Hörschelmann asks to what extent the representation of eastern Germans as young, male neo-Nazis in the western media forms part of the construction of hegemonic western masculinities and prevents, rather than enables, the contestation of international racism and neo-fascism.

The final two parts of the book focus more closely on questions of embodiment and identity. Part 4 takes a look at the ways in which men negotiate their masculine identities through bodily performances. Michael Hall analyses the relevance of sport for Australian masculine identities, while Robyn Longhurst discusses how the bodies of 'fat men' are culturally constructed as undesirable and outside of the hegemonic norm. Much recent work in feminist research has looked at the body as a canvas for the enscription of socio-cultural identities. Janine Janssen discusses this theme in relation to the very explicit example of male prisoners' tattoos. She looks at the connections between the spatial context of the prison and the men's positioning within this space as well as within society to explain the reasons for and meanings of tattooing. Her chapter is followed by Rhys Evan's auto-ethnographic reflections on the performativity of rural working-class masculinities. Rhys shows again the ambiguities and insecurities entailed in claiming masculine identity.

Part 5 concludes the book with three chapters that engage in different ways with the question of sexuality, relationships and identity. Tracey Skelton and Gill Valentine present case studies of gay boys 'coming out' and examine the ways in which this declaration alters relations within the family. They focus particularly on the negotiation of masculinity between fathers and sons and highlight the consequences of being gay for the boys' position in the family. Stuart Aitken develops this concern with men's roles in the family further by discussing different conceptions of fatherhood. He makes an empassioned plea for the recognition of fathers' emotional labour in the definition of their parental rights. Aitken's chapter shows up many of the contradictions in current conceptions of men and masculinities. He demonstrates that definitions of hegemonic masculinity hide the multiple facets of fathers' involvement in parental care and deny them significant rights. Like Aitken, Kath Browne pushes us to think more critically about our conceptions of masculinity in the final chapter. Through an analysis of the 'feminine masculinities' performed by drag kings and gay women, who are misrecognized as men, Kath illustrates a point made earlier in this introduction: that masculinities are not always and necessarily attached to male bodies. The misrecognition of women for men highlights some of the instablities of gender identity and offers an important starting point for critiquing biologically-based notions of sexuality and gender. Kath argues that 'further explorations of spatialized readings could lead to important insights into the queer disruptions and failures of gender iterations as well as the hegemonic policing that keeps dichotomous genders and sexes 'in place'. We would conclude with her that 'geographies of masculinities are well placed to explore these disjunctures alongside the hegemonic practices that maintain sexed dualisms', and hope to have inspired the reader with this book to join in this endeavour.

References

Bell, D. (2000a) 'Farm boys and wild men: rurality, masculinity and homosexuality', *Rural Sociology* 65: 547–61.

Bell, D. (2000b) 'Eroticizing the rural', in Phillips, R., Watt, D. and Shuttleton, D. (eds) *De-centring Sexualities: Politics and Representations Beyond the Metropolis*, London: Routledge.

Bell, D. and Valentine, G. (eds) (1995) *Mapping Desire: Geographies of Sexualities*, London: Routledge.

Berg, L.D. (2001) 'Masculinism, emplacement and positionality in peer review', *Professional Geographer* 53(4): 511–22.

Berg, L.D. (2002) 'Focus – Gender equity as 'boundary object': or the same old sex and power in geography all over again?', *Canadian Geographer* 46(3): 235.

Beynon, J. (2002) *Masculinities and Culture*, Buckingham: Open University Press.

Bird, I.L. (1856) *An Englishwoman in America*, London: Murray. New edition with a foreword and notes by Andrew Hill Clark published 1966, Madison: University of Wisconsin Press.

Bird, I.L. (1879) *A Lady's Life in the Rocky Mountains*, London: John Murray. New edition with an introduction by Pat Barr published in 1982, London: Virago.

Bird, I.L. (1880) *Unbeaten Tracks in Japan: An Account of Travels in the Interior, Including a Visit to the Aborigines of Yezo and the Shrine of Nikko*, New York: Putnam. New edition with an introduction by Tui Terrence Barrow published 1973, Rutland: Tuttle.

Blunt, A. and Rose, G. (1994) *Writing Women and Space: Colonial and Postcolonial Geographies*, London: The Guildford Press.

Blunt, A. and Wills, J. (2000) *Dissident Geographies: An Introduction to Radical Ideas and Practice*, Harlow: Prentice Hall.

Bly, R. (1990) *Iron John: A Book about Men*, Reading: Addison-Wesley.

Bondi, L. (1990) 'Feminism, postmodernism, and geography: space for women?', *Antipode* 22: 156–67.

Bondi, L. (1998) 'Gender, class, and urban space: public and private space in contemporary urban landscapes', *Urban Geography* 19(2): 160–85.

Brannon, R. (1976) 'The male sex role – and what it's done for us lately', in Brannon, R. and David, D. (eds) *The Forty-Nine Percent Majority*, Reading: Addison-Wesley.

Brittan, A. (1989) *Masculinity and Power*, New York: Blackwell.

Buchbinder, D. (1994) *Masculinities and Identities*, Carlton: Melbourne University Press.

Butler, J. (1990) *Gender Trouble*, London: Routledge.

Butler, J. (1993) *Bodies that Matter*, London: Routledge.

Campbell, H. and Bell, M. (2000) 'The question of rural masculinities', *Rural Sociology* 65: 532–46.

Cockburn, C. (1984) *Brothers: Male Dominance and Technological Change*, London: Pluto Press.

Collinson, D. and Hearn, J. (1996) '"Men" at "work": multiple masculinities/multiple workplaces', in Mac an Ghaill, M. (ed.) *Understanding Masculinities: Social Relations and Cultural Arenas*, Buckingham: Open University Press.

Connell, R.W. (1995) *Masculinities*, Berkeley: University of California Press.

Connell, R.W. (2000) *The Men and the Boys*, Cambridge: Polity Press.

Cornwall, A. and Lindisfarne, N. (1994) 'Dislocating masculinity: gender, power and anthropology', in Cornwall, A. and Lindisfarne, N. (eds) *Dislocating Masculinity: Comparative Ethnographies*, London and New York: Routledge.

De Pater, B. and van der Wusten, M. (1996) *Het Geografische Huis. De Opbouw van een wetenschap*, Bussum: Couthino.

Domosh, M. (1991a) 'Beyond frontiers of geographical knowledge', *Transactions of the Institute of British Geographers* 16: 488–90.

Domosh, M. (1991b) 'Toward a feminist historiography of geography', *Transactions of the Institute of British Geographers* 16: 95–104.

Edley, N. and Wetherell, M. (1996) 'Masculinity, power and identity', in Mac an Ghaill, M. (ed.) (1996) *Understanding Masculinities: Social Relations and Cultural Arenas*, Buckingham: Open University Press.

Fincher, R. and Jacobs, J.M. (1998) *Cities of Difference*, New York: Guildford Press.

Halberstam, J. (1998) *Female Masculinity*, Durham: Duke University Press.

Halberstam, J. (2002) 'An Introduction to Female Masculinity: Masculinity without Men', in Adams, R. and Savran, D. (eds) *The Masculinity Studies Reader*, Oxford: Blackwell.

Hall, S. (1997) 'The Spectacle of the "Other"', in Hall, S. (ed.) *Representation: Cultural Representations and Signifying Practices*, London: Sage.

Halsall, P. (1998) 'Ancient History Sourcebook: Documents on Greek Slavery, c. 750–330 BCE'. Online. Available at http://www.fordham.edu/halsal/ancient/greel-slaves.html (accessed 20 February 2004).

Hearn, J. (1996) 'Is masculinity dead? A critique of the concept of masculinity/masculinities', in Mac an Ghaill, M. (ed.) *Understanding Masculinities: Social Relations and Cultural Arenas*, Buckingham: Open University Press.

Jackson, P. (1994) 'Black male: advertising and the cultural politics of masculinity', *Gender, Place and Culture* 1(1): 49–59.

Jackson, P., Stevenson, N. and Brooks, K. (2001) *Making Sense of Men's Magazines*, Cambridge, Polity Press.

Kimmel, M. (1994) 'Masculinity as homophobia: fear, shame and silence in the construction of gender identity', in Brod, H. and Kaufman, M. (eds) *Theorizing Masculinities*, Thousand Oaks, London, New Delhi: Sage Publications.

Knopp, L. (1994) 'Social justice, sexuality and the city', *Urban Geography* 15(7): 644–60.

Kosofsky Sedgwick, E. (1995) ' "Gosh, Boy George, you must be awfully insecure in your masculinity" ', in Berger, M., Wallis, B. and Watson, S. (eds) *Constructing Masculinities*, New York: Routledge.

Laurie, N., Dwyer, C., Holloway, S. and Smith, F. (1999) *Geographies of New Femininities*, Harlow: Longman.

Liepins, R. (2000) 'Making men: the construction and representation of agriculture-based masculinities in Australia and New Zealand', *Rural Sociology* 65: 605–20.

Little, J. (2002) 'Rural geography: rural gender identity and the performance of masculinity and femininity in the countryside', *Progress in Human Geography* 26(5): 665–70.

Little, J. and Panelli, R. (2003) 'Gender research in rural geography', *Gender, Place and Culture* 10(3): 281–90.

Longhurst, R. (2000) 'Geography and gender: masculinities, male identity and men', *Progress in Human Geography* 24(3): 439–44.

Longhurst, R. (2001) *Bodies. Exploring Fluid Boundaries*, London and New York: Routledge.

Mac an Ghaill, M. (1994) *The Making of Men: Masculinities, Sexualities and Schooling*, Buckingham: Open University Press.

Mac an Ghaill, M. (ed.) (1996) *Understanding Masculinities: Social Relations and Cultural Arenas*, Buckingham: Open University Press.

McDowell, L. (1997) *Capital Culture: Gender at Work in the City*, Oxford: Blackwell.

McDowell, L. (1998) *Gender, Identity and Place: Understanding Feminist Geographies*, Cambridge: Polity Press.

McDowell, L. (2001) 'Men, management and multiple masculinities in organisations' *Geoforum* 32(2): 181–98.

McDowell, L. (2002) 'Transitions to work: masculine identities, youth inequality and labour market change', *Gender Place and Culture* 9(1): 39–59.

McDowell, L. (2003) *Redundant Masculinities: Employment Change and White Working Class Youth*, Oxford: Blackwell.

MacInnes, J. (1998) *The End of Masculinity. The Confusion of Sexual Genesis and Sexual Difference in Modern Society*, Buckingham and Philadelphia: Open University Press.

Massey, D. (1994) *Space, Place and Gender*, Cambridge: Polity Press.

Momsen, J. Henshall (1991) *Women and Development in the Third World*, London: Routledge.

Momsen, J. Henshall (1993) *Different Places, Different Voices: Gender and Development in Africa, Asia and Latin America*, London: Routledge.

Monk, J. and Hanson, S. (1982) 'On not excluding the other half of the human in Human Geography', *Professional Geographer* 34(11): 11–23.

Morgan, D. (1992) *Discovering Men*, London: Routledge.

Morgan, D. (2001) 'Family, gender and masculinities', in Whitehead, S.M. and Barrett, F.J. (eds) *The Masculinities Reader*, Cambridge and Oxford: Polity Press.

Mort, F. (1996) *Cultures of Consumption: Masculinities and Social Space in Late Twentieth-Century Britain*, London and New York: Routledge.

Mosse, G.L. (1996) *The Image of Man: The Creation of Modern Masculinity*, Oxford: Oxford University Press.

Pain, R. (2001) 'Gender, race, age and fear in the city', *Urban Studies* 38(5/6): 899–914.

Peake, L. (1993) ' "Race" and sexuality: challenging the patriarchal structuring of urban social space', *Environment and Planning, Part D: Society and Space* 11(4): 415–33.

Peart, K. (1998) 'Victorian Arts & Crafts and Pre-Raphaelite', updated intranet site produced for Masters in Art – Computing and Art History, Birkbeck College, London University. Online. Available at http://freespace.virgin.net/k.peart/Victorian/millnwpass.htm (accessed 20 February 2004).

Phillips, R. (1997) *Mapping Men and Empire: A Geography of Adventure*, London and New York: Routledge.

Radcliffe, S.A. (1998) *Gender in the Third World: A Geographical Bibliography of Recent Work*, Institute of Development Studies at the University of Sussex.

Rose, G. (1993) *Feminism and Geography: The Limits of Geographical Knowledge*, Cambridge: Polity Press.

Rutherford, J. (1997) *Forever England: Reflections on Masculinity and Empire*, London: Lawrence and Wishart.

Saugeres, L. (2002) 'Of tractors and men: masculinity, technology and power in a French farming community', *Sociologia Ruralis* 42: 143–59.

Segal, L. (1990) *Slow Motion: Changing Masculinities, Changing Men*, London: Virago.

Seidler, V.J. (1989) *Rediscovering Masculinity: Reason, Language and Sexuality*, London: Routledge.

Staples, R. (1982) *Black Masculinity: The Black Male's Role in American Society*, San Francisco: The Black Scholar Press.

Stoddart, D.R. (1991) 'Do we need a feminist historiography – and if so, what should it be?', *Transactions of the Institute of British Geographers* 16: 484–7.

Walby, S. (1990) *Theorizing Patriarchy*, Oxford: Blackwell.

Whatmore, S., Marsden, T., Lowe, P. and Clout, H. (1994) *Gender and Rurality*, London: Fulton.

Whitehead, S.M. and Barrett, F.J. (eds) (2001a) *The Masculinities Reader*, Cambridge and Oxford: Polity Press.

Whitehead, S.M. and Barrett, F.J. (2001b) 'The sociology of masculinity', in Whitehead, S.M. and Barrett, F.J. (eds) *The Masculinities Reader*, Cambridge and Oxford: Polity Press.

Willis, P. (1977) *Learning to Labour: How Working Class Kids Get Working Class Jobs*, London: Saxon House.

Winchester, H.P.M. (1999) 'Interviews and questionnaires as mixed methods in population geography: the case of lone fathers in Newcastle, Australia', *Professional Geographer* 51(1): 60–7.

Women and Geography Study Group (1984) *Geography and Gender*, London: Hutchinson.

Zelinsky, W., Monk, J. and Hanson, S. (1982) 'Women and geography: a review and prospectus', *Progress in Human Geography* 6: 317–66.

Part 1

MASCULINITIES IN TRANSITION

2

THE MEN AND THE BOYS

Bankers, burger makers and barmen

Linda McDowell

Summary

In this chapter I examine the interconnections between economic restructuring, the rise of service sector occupations and the social construction of masculinity. Through case studies of different forms of service work – high and low status – I show how embodied performances that emphasize what are traditionally seen as characteristics of femininity are now a key part of many of the service occupations undertaken by men. Although both high status and low status occupations demand new forms of workplace performances from male employees, significant inequalities between these types of work remain. Young working-class men, for whom the type of manufacturing work that their fathers might have entered has now disappeared, find it increasingly difficult to gain access to the sort of work that will enable them to establish independent living and adopt the traditional breadwinner role.

- The material and discursive construction of masculine identities
- Polarization in the service sector
- Doing embodiment: high tech/high status work in merchant banks
- Learning to serve: embodied performances in bars, cafés and fast food outlets.

Introduction: the material and discursive construction of masculine identities

In the last decade there has been a rapid growth in the scholarly literature about masculinity (see, for example, Whitehead's (2002) review and the reader by Whitehead and Barrett (2001)), paralleling earlier debates about the multiple construction of femininity, its diversity and mutability, and challenging singular notions of masculine identity. Until recently, however, the association of masculinity with the sphere of waged work in capitalist societies has been taken for granted. The very definition of hegemonic masculinity in industrial societies is bound up with labour market participation. Being a real man involves paid employment, whether in the embodied spaces of manual labour or the cerebral spheres of high-tech industry, business services or science. This association between men and the labour market has been so dominant that until relatively recently the complexities and changing nature of the association has been under-theorized. It has been women's exclusion from the public spaces of waged employment, as well as their segregation into a particular range of occupations identified as appropriately feminine, that has been the key focus of theorizing by second-wave feminists. However, in more recent analyses, the variety in men's association with employment has become the focus of theoretical attention. The significance of a dual construction of masculinity – a disembodied, rational, bureaucratic and scientific middle-class masculinity and a more vital, embodied and less cerebral masculinity based on male strength associated with manual occupations and working-class men – has been analysed and illustrated in a range of empirical case studies of different types of masculinity in different forms of employment or occupations. As Connell (1995) noted, 'definitions of masculinity are enmeshed in the history of institutions and economic structures. Masculinity is not just an idea in the head or a personal identity' (p. 29). Further relations of power *between* men, as well as between men and women, have been documented. Like femininity, masculinity is a multiple social construction in which men are positioned in relation to each other, often in terms of inferiority and superiority.

While Connell's insistence that masculinity is enmeshed in economic structures was a crucial step in understanding the links between masculinity and employment, more recent analysts have refocused attention from institutions to social identities, on the ways in which masculine identity is constructed by individuals and by groups of men in their heads as it were through their positioning within a whole range of different discourses and social practices, including literature, music and other forms of popular and high culture that portray idealized versions of desirable masculinity, as well as through the operation of the rules and regulations of a wide range of social institutions and men's participation in their practices. In capitalist societies the construction of masculinities occurs in a range of institutions and spaces from the family to the school, on the football field and its terraces, as well as in the workplace. The significance of these different institutions varies, not only across the life cycle of individual men but also in different historical circumstances. In his more recent work, for example, Connell (2000) has noted that 'the institutions of competitive sport seem peculiarly important for contemporary western masculinities' (p. 11). Indeed, sport as a leisure activity has become elided with working life for growing numbers of people, both men and women, for whom a fit body is both a key attribute of successful employment and of hegemonic versions of masculinity/femininity. Further elision occurs in expanding numbers of high status workplaces where sports facilities are provided on the premises for employees to maximize their efforts and minimize time away from the workplace.

This current emphasis on sporting prowess makes abundantly clear the recognition that the body is a key element in social construction of gendered identities (Featherstone *et al.* 1991; Teather 1999). Men's and women's bodies are surfaces that are inscribed with, defined by and disciplined through social norms and conventions about gendered appearances, in size, weight and deportment as well as through decoration and clothing (Bordo 1993; Grosz 1994): all of which are features that have become increasingly important in the new forms of service-based work that dominate industrial economies. Thus, through bodily performances, as well as in all social interactions, masculinities (and femininities) are constantly being actively constructed, maintained or challenged in the different spaces of daily life. As West and Zimmerman (1987) noted some time ago now, gender is a performance, an active construction, people 'do gender' as an everyday act (Butler 1990; Kondo 1990). Thus it is now widely recognized that there are multiple ways of being a man; masculinities are complex and often contradictory, and may be riven with conflicting desires and ambivalences. Even that version of masculinity currently defined as dominant through a wide range of practices – heterosexuality, labour market participation, a fit and young, preferably white, body – has to be constantly re-created and policed and regulated. As Mac an Ghaill (1994, 1996) has argued, hegemonic masculinity (the most highly valued version of being a man) is 'constituted by cultural elements which consist of contradictory forms of compulsory heterosexuality, misogyny and homophobia. These are marked by ambivalence and contingency' (1996: 133). Men continually (re)define themselves as something that women – or less highly valued versions of masculinity – are not; masculinity is constructed in opposition to sets of characteristics that are regarded as inferior. Thus, in his empirical work with young men, Mac an Ghaill has noted that 'what emerges as of particular salience is the way in which heterosexual young men are involved in a double relationship: of disparaging the 'other', including women and, and at the same time, expelling femininity and homosexuality from within themselves' (1996: 133). For men in the new workspaces of the service sector, there are complex negotiations in the construction of acceptable versions of masculinity, especially in workplaces that seem to value so-called feminine attributes of empathy and caring, docility and deference.

Polarization in the service sector

In the rest of this chapter I want to bring together this theoretical work about multiple masculinities, the claims about a growing crisis of masculinity and new forms and patterns of employment participation, examining the ways in which the connections between labour market change and masculine identities are changing in contemporary Britain. I shall illustrate my argument through two case studies of different groups of men and particular occupations in the service sector. Here, I shall draw on my own recent work (McDowell 2002a, 2002b, 2003) with young working men just entering the labour market in a period when they are increasingly portrayed as failures compared with young women of their age, as well as returning to my earlier work in merchant banks (McDowell 1997). As is now well established in western capitalist economies, service sector occupations have expanded as manufacturing employment has declined from its peak (in Britain at least in the mid-1950s when more than half of all employees were working in manufacturing industries). Currently in 2001 in Britain, for example, over two-thirds of the labour force is employed in the service sector. As commentators have suggested, the services sector labour market is becoming increasingly polarized, in which two distinctive groups of workers labour under different conditions and for differential rewards. At the top end of the hierarchy, highly educated men and increasingly women, work for the legal firms, business services and banks that have proliferated in global cities, as well as in the high tech sector of knowledge-based industries including dot.com companies (though fewer of these than there were), hardware and software developers, and the electronics industry in which innovation, flexibility and mobility are the highly valued attributes of the so-called 'information age' (Castells 2000a), or the 'new economy' (Carnoy 2000). These are the employees who are benefiting from the risk society' (Beck 1992), whose ability to capitalize on their individual characteristics leads to the successful development of a 'portfolio' career, dominated by movement between employers as new projects arise and whose reward for relative insecurity is high annual pay, often supplemented by performance bonuses.

At the opposite end of the service labour market some of the same characteristics of flexibility, mobility and insecurity are also common attributes but here there is little associated recompense for the risks of impermanent employment and short-term contracts. Instead, low wage work in what might be termed 'servicing' jobs – the fast food, retail sector, bar work, messenger services, sandwich shops, and coffee outlets that have expanded so rapidly in British towns and cities – as well as the tourist and hospitality industry and the now often privatized local and central government community services, is based on the rapid turnover of predominantly young workers often employed on a less than full-time basis on temporary contracts. Manuel Castells (2000b), an analyst of the new information age, has termed these two distinctive forms of labour 'networked' or 'programmable labour' (the educated, flexible and highly motivated workers in the high tech sector) and 'generic labour' (the less educated, unskilled and transferable or substitutable labour of what Brush (1999) has defined as the 'high touch' sector). What unites both types of worker, however, is the significance of a bodily performance, in which the personal attributes of an employee are a key part of the service to be exchanged, whether it is financial advice of the most arcane nature or the purchase of a cappuccino in a city centre coffee shop. The new service-based economy requires that employees embody what are regarded as desirable social attributes, including weight, skin colour, stance and style in order that the interactions between service providers and consumers are productive and profitable.

This polarization of labour and the common emphasis on embodied performances is partly paralleled in the built environment as new spaces of work are created to replace the older industrial landscape. Thus the post-modern spaces of new financial business districts reflect 'the imperatives of a post-industrial economy, of the internationalization of fictive commodities, of financial business and culture services'

(Lash 1990: 72) where the programmable labourers work in spaces that seem to elide work and leisure. Here bankers, business and software developers work in glass boxes and towers, with plant-filled central atria. In redeveloped central areas new office developments may be associated with retail and leisure spaces, including upmarket clothes shops, and as I noted earlier with gymnasiums and other sports' provision. Elsewhere in the city, as well as within the central business districts, generic labourers provide services, whether to the masses or the financial elite – Tom Wolfe's (1990) 'masters of the universe' – in post-modern playful reconstructions of historical Italian hill villages reborn as shopping malls or in the more industrial landscapes of fast-food outlets and retail chains. In these spaces, a 'lifestyle' is available for purchase from workers whose performance writes consumers into the script as pseudo-friends and acquaintances. Thus, on entering the US clothes retailer the Gap, for example, a friendly young person greets customers, asking with apparent interest how their day has gone so far. In both types of service spaces, men perform masculinity in ways that both build on but also challenge their sense of themselves as masculine. In the next two sections, I look first at the workplace performances of male middle-class bankers and then at working-class fast food employees, showing how, in these different workspaces, young men tend to emphasize those aspects of their working lives that are congruent with themselves as men but are also confronted with decisions about how to manage their bodies and appearances which are now an integral aspect of a successful workplace performance. Thus even in the still highly-masculinized spaces of banking, what might be regarded as typically feminine attributes such as care of the body or deference to clients are now a key part of many men's daily working lives.

Doing embodiment: high tech/high status work in merchant banks

In my earlier work (McDowell 1997, 2001) with employees of merchant banks in the City of London, I argued that new ways of performing masculinity were becoming evident in the high status offices and board rooms of corporate banking. I contrasted these spaces with those of the trading floor and dealing rooms, spaces that have become familiar through popular representations of merchant banking in films such as *Trading Places*, *Wall Street* and *Rogue Trader*. Here I want to recap the arguments of my book, *Capital Culture*, showing how, in both working spaces in banking, different forms of masculine performances take precedence, but which in both spaces emphasize particular aspects of embodiment and which also retain and reinforce the older patterns of male dominance in the workplace.

I want first to consider the high tech work of trading and dealing and its association with a loud, aggressive, macho version of masculinity. In many ways since I undertook the empirical work on which *Capital Culture* was based, innovations, especially in screen-based trading, have largely replaced the old carnivalesque embodied world of the trading floor that used to be dominant. The early 1990s saw the final

expression of an older version of doing gender in banks that has now been transformed into a more modest set of behaviours in which fast thinking and the rapid
manipulation of computer keyboards have replaced the fast, frenzied sweating,
jostling crowd of aggressive young men screaming prices and bids across the heads of
the mass of people on the trading floor. In this earlier manifestation of masculine
dominance, women were excluded by their (lack of) physical presence, their generally smaller size and higher voices which made it hard for them to be seen and heard
in the crowd. As I argued then:

> Stock exchanges are sites of spectacle where exotic goods are exchanged
> and the body is allowed out of control – shouting and gesticulating are
> required forms of interaction. Like a medieval fair, the exchange is a dual
> site – a market place, a site of commercial exchanges . . . but also a site of
> pleasure, unconnected to the real world and standards of normal behavi-
> our. On the trading floors . . . performances are stimulated by excitement
> and desire. The usual control of emotions expected in middle-class work-
> places is not demanded.
>
> (McDowell 1997: 167)

This world is one that the former Salomon Bros employee, Michael Lewis
(1989), termed 'a jungle' in which the loudest, most aggressive, outrageously heterosexual male is the king. And as numerous merchant banks have recently found to
their detriment in a series of costly sexual harassment cases brought in the first two
years of the new millennium, it is a world in which discrimination against women is
taken for granted. Even in the newer, less physical, world of screen-based trading,
types of behaviour that would not be tolerated in most workspaces persist. Loud
shouting on the telephone, high jinks in the office, practical jokes and physical horseplay in which the exuberant high spirits of the predominantly young male employees
find release from the stresses of monitoring the movements of the money market
continue to be common forms of behaviour. It is a high risk and extremely
competitive culture. As Ryle (2002) noted in a recent article about court cases
involving the banking sector, 'harassment cases tend to centre on remarks about
skirt lengths, looks or lewdness' (p. 5). Deutsche Bank, for example, paid out
£70,000 in compensation in 2001 to a woman employee who was habitually
described as a 'bit of skirt' and as 'hot totty'. But it is not only women who are
uncomfortable in this environment. Men whose masculinity is not based around the
loud performance of aggressive heterosexual masculinity also feel out of place, and
occasionally take cases of harassment and discrimination to the courts.

The sort of embodied, high spirited masculine performance that continues to
dominate the trading and dealing arenas in merchant banks is, however, inappropriate and out of place in the offices and the board rooms of corporate finance sections
that comprises the 'other side' of merchant banking. Here concerns with weight,
accent and clothes, the production of a particular version of the dominant or most

highly valorized masculinity characterized by a tight, trim, white, middle-class body appropriately garbed in dark suits and crisp white shirts, is what is required by the employer. The key role of employees in this arena of merchant banking is to meet clients and convince them that these young bankers have access to superior knowledge and specialized advice in the take-overs, mergers and acquisitions that brings high profits to the bank. Here young men in the main, but increasingly young women, from the right background, the right school and with a good degree from a good university – all essentially middle-class attributes in contemporary Britain – are recruited to persuade clients that they (the bankers) know best. Thus as interviewees told me 'we have only one thing to sell, and that is ourselves'. 'Merchant bankers have to be presentable and come in and convince' or as another young banker explained at greater length 'you have to know the right people and be convincing. So really the ability to get on with people and go out and sell is essential'. In this world, employees whose social characteristics most closely parallel those of the clients tend to have the advantage. In consequence, corporate finance remains a largely middle-class world, dominated in its higher echelons by men. The younger men whom I interviewed emphasized the need to work hard on their personal appearance through attention to clothes, weight, style, and the body in general as well as to construct a carefully judged performance that emphasizes both their technical superiority and their class similarities, as well as cultivating an empathetic and often deferential manner towards their clients. Thus in some senses, their personal performance combines stereotypical masculine and feminine characteristics as it includes a range of attitudes and behaviours that have tended to be designated as typically feminine. Indeed, a number of women bankers whom I interviewed remarked rather caustically that there was no way that they could challenge this apparently winning combination of feminine attributes in a masculine body: what Suzanne Moore (1988) memorably dubbed, in a different context, 'the pimps of postmodernism' who were 'getting a bit of the other'.

Learning to serve: embodied performances in bars, cafés and fast food outlets

My second example moves from the high tech to the high touch workers, from the programmable to the generic labourers of a post-industrial service society. While the working lives of middle class male corporate bankers might seem a world away from the lives of young male employees of the fast food outlets, coffee houses and department stores that increasingly dominate the central spaces of British towns and cities, in many ways there are several similarities in the emphasis on an individualized, embodied workplace performance. Here too young men have nothing to sell but themselves in their search for employment and (much lower) financial rewards, and weight, style, clothes and accent also play a part in their success. In a more recent piece of research (McDowell 2003), I have been talking to young white male school leavers from working class families in Cambridge and Sheffield in an attempt to understand their sense of themselves as masculine as they search for work after leaving school.

In Sheffield, as in many deindustrializing cities, young men with few educational credentials and little social capital face a different world than their fathers did on leaving school between two and three decades earlier. The old gender division of labour in which working-class men accepted life-time toil in a heavy industry in return for a steady 'breadwinners' wage that enabled them to reach at least a decent standard of living and to support a family has vanished. Instead part-time labour, casual work and low pay have become a more common working pattern for many young people when they enter the labour market. As employment opportunities in offices, shopping centres and in the hospitality industry increasingly replace manufacturing employment, women's labour market participation rates have risen. In many of the more affluent small towns in the UK, especially in the buoyant labour markets of the south-east, in Guildford, Winchester and Brighton for example, women now outnumber men as employees, not just in the service sector but in the local economy as a whole. In low wage service employment, women are often the preferred employees as their social skills and bodily presence more accurately reflect the attributes that employers desire. Boys are messier and less likely to have educational credentials, more likely to find the work boring and to start to play around, less willing to defer to the customer who must be greeted pleasantly and is always, always right (Newman 1999). Thus they may find it harder to secure and to hold on to the type of employment that often is all that is available both in deindustrializing towns in northern and peripheral regions as well as in the market towns in the south. In a study of young men in the US, Philippe Bourgois (1995) noted that the macho bravado and protest masculinity that young urban men adopt on the streets of New York City disqualified them from many job vacancies. Similar young men in Britain, surly and aggressive, often tattooed and pierced, do not fit into the ambience of an 'Italian' café, an up-market winebar or even the industrialized spaces of a fast food outlet where hard work, discipline and a deferential workplace performance are essential requirements for continuous employment. Unable to accept that the economy has changed or to adapt their behaviour, many young men blame the 'other' – usually women or 'immigrants' – for their lack of labour market success and tend to exaggerate the very qualities in their own behaviour that first disqualified them from employment (Fine and Weiss 1998).

In Sheffield I followed a group of young white men for 18 months between April 1999 and September 2000 as they left school with few or no qualifications and began to search for work (see McDowell 2000, 2002a, 2003). I also contacted them again two years later in September 2002. Typically, their first jobs were in shops, bars and, in particular, in Sheffield in a branch of McDonalds in a nearby shopping mall. A small number of the group obtained types of work that they regarded as more suitable for them as men, congruent with their notions of acceptable youthful masculinity, in, for example, a garage, a woodcutting firm or a small steel firm. Here, however, I want to focus on the experiences of the young men who found jobs in the bottom end of the service sector. Richard, for example, worked in McDonalds for a few months between two brief spells in the retail sector (first at Sports' Soccer and then at Burtons (a men's outfitters)) followed by work as a window cleaner. At Sports' Soccer, he told me that the work was reasonable and that it 'suited a man, like, cos it's about sport' but the work was monotonous. Richard found Burtons too staid and too poorly paid, and so he left almost immediately moving to McDonalds, a few yards away in the same shopping centre, where the hourly rate of pay was higher. Two other members of his year group also worked at McDonalds and all three of them found the work relentless and unremitting as well as a challenge to their developing sense of self. The flexible shifts that were required were difficult to deal with, for example, as Richard explained:

> Sometimes they like me on early, but at 6am when I have to get up I'm really hurting. It's very bad, right annoying. I should work 6 til 2, but it can be 3, even 4 when I finish. And then you get changed around. Sometimes 10 to 5 or 11 to 6, or then you do a closing; start at 5 and work til it closes.

As well as flexible hours which made it difficult for young men not until then notable for their punctuality or promptness to acquire the necessary workplace discipline and routine, the work itself was both demanding and demeaning. As Richard vividly described:

> They watch you; there's people watching you work all the time, passing information onto the manager. And they push people around, management does that . . . People don't last long here. They leave after about two months because they can't handle it. It's like slave labour.

The work in a fast food outlet is a classic Fordist labour process (Gabriel 1988; Schlosser 2001), divided into individual tasks, each undertaken by a different worker. Here is Richard's description: 'we work in a circle as a team. So to do like a Big Mac or something, there's one on table dresses the Big Mac, one passes the bun onto grill and another takes it off so it's right quick'. The employees must perform each action in a specific time, as well as greet customers with a scripted set of remarks, and bring to work each day a clean pressed uniform so that they present a neat and conformist appearance.

But perhaps the most difficult part of the requirements in this type of job is dealing with customers, many of whom are the same age and from the same backgrounds as the workers, if not, indeed, their former classmates and friends. Fast food jobs are among the most stigmatized and denigrated of all service sector work. Indeed, the term 'McJob' is used as a category to summarize all the worst attributes of the bottom end of the service sector, imbued with stigma and lack of the values and moral worth that characterized the former Protestant ethic in that labouring jobs in the manufacturing sector that have now largely disappeared (Bauman 1998). For fast food employees it is hard to maintain their dignity or to retain their self-respect in a front line job where customers often treat the employees as worthless. Many customers are rude, demanding and insulting and yet workers must respond with a smile. As Katherine Newman (1999) has argued in her study of a fast food outlet in New York city, 'workers must stifle their outrage [and] tolerate comments that would almost certainly provoke a fight outside the workplace' (p. 89). Perhaps the most difficult part of this deferential performance is learning to deal with peers, who may not only be rude and insulting but also expect favours from their former school friends or neighbours. Richard told me how hard he found it to refuse 'mates who come in and want a free burger or fries'. Similarly, John who found work in a city centre bar, at first clearing tables but then serving drinks, found himself plagued by friends looking for a free drink: 'I were sacked in the end; the manager said that too many of me mates came in.'

In these types of occupations, however, the continuing dominance of masculinity and masculine attributes that construct women as inferior 'others' in high status occupations such as merchant banking is not evident. Instead, both young men and young women, who work together doing similar tasks, are constructed as disposable and replaceable labour in jobs where rates of turnover are exceptionally high and workplace loyalty and career progression are not expected by either the employers or the employees.

Conclusions

In both the high status spaces of merchant banking and the low status spaces of fast food outlets and bars, new versions of masculinity are required workplace performances. The old binary distinction between a cerebral, rational middle-class masculinity that was essentially disembodied and a working-class, hard, manual, powerfully embodied masculinity (both of which were regarded as 'natural' and needing little explicit construction or attention to the details of personal interaction or bodily presentation) has been replaced by new versions of masculinity in a service-dominated economy. In this new world of work, care of the customer, empathy, with both clients and co-workers, and a scripted embodied performance emphasizing, in different contexts, clothes, weight, and cleanliness, as well as humility and deference, or at least respect of others' wishes and demands, are essential employment attributes. These shifts have both challenged the automatic privileges of masculinity and have required new forms of masculine presentation and performance in attaining and retaining employment in a more uncertain world where the old notion of employment for life by a single employer, at least for the majority of men, has disappeared. Whether this transformation has begun to alter the gender balance of privilege in the workplace remains an as yet unanswered question. Despite the apparent 'feminization' of workplace performances, men of all ages continue to earn more than women in similar occupations. Even in the bottom end service jobs, for example, a gender pay gap is already apparent among young workers by the age of 19. It also clear, however, that even if gender divisions might be changing, patterns of class inequality are becoming ever more firmly cemented in advanced industrial economies. The growing polarization between workers, exemplified here by the comparison between high status bankers and low status fast food workers, is paralleled by a growing gap in income levels in Britain, in other western European nations and, in particular, in the US in the new millennium. There is a real prospect that the poorest workers in society will be permanently excluded from attaining the life chances and opportunities expected and taken for granted by the majority. For the young men like Richard and John whose voices are represented here, the prospect of attaining the correlates of adult masculinity associated in the past with a male breadwinner status and wage is increasingly less likely than in earlier decades. The affluent lifestyle open to the young bankers also heard here is not even part of their hopes for the future.

References

Bauman, Z. (1998) *Work, Consumerism and the New Poor*, Buckingham: Open University Press.

Beck, U. (1992) *The Risk Society: Towards a New Modernity*, London: Sage Publications.

Bordo, S. (1993) *Unbearable Weight: Feminism, Western Culture and the Body*, Berkeley and Los Angeles: University of California Press.

Bourgois, P. (1995) *In Search of Respect: Selling Crack in El Barrio*, Cambridge: Cambridge University Press.

Brush, L. (1999) 'Gender work, who cares?! Production, reproduction, deindustrialization and business as usual', in Ferree, M.M., Lober, J. and Hess, B. (eds) *Revisioning Gender*, London: Sage Publications.

Butler, J. (1990) *Gender Trouble*, London: Routledge.

Carnoy, M. (2000) *Sustaining the New Economy: Work, Family and Community in the Information Age*, New York: Russell Sage Foundation; Cambridge, Mass: Harvard University Press.

Castells, M. (2000a) *The Information Age: Economy, Society and Culture* (updated edition, 3 volumes), Oxford: Blackwell.

Castells, M. (2000b) Materials for an exploratory theory of the network society, *British Journal of Sociology* 51: 5–24.

Connell, R.W. (1995) *Masculinities*, Cambridge: Polity.

Connell, R.W. (2000) *The Men and the Boys*, Cambridge: Polity.

Featherstone, M., Hepworth, M. and Turner, B. (eds) (1991) *The Body: Social Process and Cultural Theory*, London: Sage Publications.

Fine, M. and Weiss, L. (1998) *The Unknown City: The Lives of Poor and Working Class Young Adults*, Boston: Beacon Press.

Gabriel, Y. (1988) *Working Lives in Catering*, London: Routledge.

Grosz, E. (1994) *Volatile Bodies*, Bloomington: Indiana University Press.

Kondo, D. (1990) *Crafting Selves: Power, Gender and Discourse of Identity in a Japanese Workplace*, Chicago: University of Chicago Press.

Lash, S. (1990) 'Postmodernism as humanism?', in Turner, B. (ed.) *Theories of Modernity and Postmodernity*, London: Sage Publications.

Lewis, M. (1989) *Liar's Poker: Two Cities, True Greed*, London: Hodder and Stoughton.

Mac an Ghaill, M. (1994) *The Making of Men: Masculinities, Sexualities and Schooling*, Buckingham: Open University Press.

Mac an Ghaill, M. (ed.) (1996) *Understanding Masculinities: Social Relations and Cultural Arenas*, Buckingham: Open University Press.

McDowell, L. (1997) *Capital Culture: Gender at Work in the City*, Oxford: Blackwell.

McDowell, L. (2000) 'Learning to serve? Employment aspirations and attitudes of young working-class men in an era of labour market restructuring', *Gender, Place and Culture* 7: 389–416.

McDowell, L. (2001) 'Men, management and multiple masculinities in organisations', *Geoforum* 32(2): 181–98.

McDowell, L. (2002a) 'Transitions to work: masculine identities, youth inequalities and labour market change', *Gender, Place and Culture* 9: 39–59.

McDowell, L. (2002b) 'Masculine discourses and dissonances: strutting "lads", protest masculinity and domestic respectability', *Environment and Planning D: Society and Space* 20: 97–119.

McDowell, L. (2003) *Redundant Masculinities? Employment Change and White Working Class Youth*, Oxford: Blackwell.

Moore, S. (1988) 'The pimps of postmodernism: getting a bit of the other', in Chapman, R. and Rutherford, J. (eds) *Male Order: Unwrapping Masculinity*, London: Lawrence and Wishart.

Newman, K. (1999) *No Shame in my Game: The Working Poor in the Inner City*, New York: Russell Sage Foundation and Knopf.

Ryle, S. (2002) 'Court cases pile up as sex and the city collide', *The Observer Business*, 13 October, p. 5.

Schlosser, E. (2001) *Fast Food Nation: What the All-American Meal is Doing to the World*, London: Allen Lane.

Teather, E. (1999) *Embodied Geographies: Spaces, Bodies and Rites of Passage*, London: Routledge.

West, C. and Zimmerman, D. (1987) 'Doing gender', *Gender and Society* 1: 125–51.

Whitehead, S. (2002) *Men and Masculinities*, Cambridge: Polity.

Whitehead, S. and Barrett, F. (eds) (2001) *The Masculinities Reader*, Cambridge: Polity Press.

Wolfe, T. (1990) *The Bonfire of the Vanities*, London: Picadore.

Further reading

Connell, R.W. (1995) *Masculinities*, Cambridge: Polity. The best place to start exploring the growing literature on masculinity is with the social theorist R.W. Connell. His book *Masculinities* is now a classic in which he established a theoretical framework for understanding the social organization of masculinity and its history, illustrating his arguments with four case studies from his own empirical work.

Connell, R.W. (2000) *The Men and the Boys*, Cambridge: Polity. Connell's more recent book (the title was ideal for my chapter and as it is a version of a popular saying I decided I might use it too, albeit with a subtitle) is a review of a wide range of work and reflects the discursive shift in theorizing about gender.

Whitehead, S. (2002) *Men and Masculinities*, Cambridge: Polity. *Men and Masculinities* also provides a useful review of alternative theoretical approaches to the analysis of masculinity, as well as an interesting discussion of men's relationship to feminism.

3

THE LOGIC AND LIMITATIONS OF MALE WORKING-CLASS CULTURE IN A RESOURCE HINTERLAND

Thomas Dunk and David Bartol

Summary

The example discussed in this chapter focuses on working-class masculinity in a North American industrial 'hinterland' region. The region became known as an area of working-class radicalism in the first several decades of the twentieth century, and in the 1950s, 1960s, and 1970s, the industrial base provided relatively secure and well-paid employment for male, blue-collar workers in an era of industrial decline stimulated by global and local environmental and economic problems. This chapter will describe the interactions between these concomitant processes of formation and dissolution of masculinist identities and their changing political and social significance.

- Masculinity, class, space, and time
- Regional context and history
- Masculinity, class, and regional identity during the Fordist era
- Post-Fordist realities and working-class culture.

Masculinity, class, space, and time

Connell (1995) has shown that modern societies contain a variety of forms of masculinity that are ordered hierarchically. What he terms hegemonic, subordinate, complicit and marginal masculinities (Connell 1995) differ in terms of their relationships with the femininities and social classes of a given society. Pyke (1996) argues that class is the form of power that effectively locks the different types of masculinities into dominant or subordinate positions:

> the ascendant masculinity of higher-class men and the subordinate masculinity associated with lower-class men are constructed in relation to one another in a class-based gender system. Class-based masculinities provide

men with different mechanisms of interpersonal power that, when practiced, (re)constitute and validate dominant and subordinate masculinities.

(pp. 527–8)

Differences between the forms in which masculinity is expressed are related to the material realities of different class positions. Given the constraints on life chances and the nature of the work in which they are involved, it is not surprising that working-class men tend to celebrate physical toughness and embrace traits of machismo. Men of the new middle class, on the other hand, frequently display more competitive attitudes and concentrate on upward mobility and success. Of course, both working-class and middle-class men define themselves at least partly in relationship to women, as, for example, in the ideology of the breadwinner for the family (Wright 2001).

Given that classes are not distributed evenly across the globe or within nations, particular expressions of masculine and class identity are often bound up with regional identities as well. Conflicts between forms of masculinity, between genders, and between classes are inextricably connected to spatial identities and conflicts. And as the nature of work and the forms of relationships and behaviour that are required of workers change, ideas about appropriate or normal gender identities change as well. For example, de-industrialization places men socialized into the masculine culture of the industrial shopfloor in a challenging position. If they are required to seek employment in service-sector positions they discover that opportunities for masculine expression and the modes of expression considered appropriate are different than those to which they are accustomed. Thus, forms of masculinity must be understood in relationship to the ever-changing set of class, spatial, and temporal relationships in which actual living subjects are enmeshed. This chapter explores these relationships through an examination of male, working-class culture in northwestern Ontario in Central Canada (see Figure 3.1).

Regional context and history

Northwestern Ontario comprises of the three westernmost districts of the central Canadian province. In terms of its physical geography, it is a harsh region. Winters are typically long and cold and summers are relatively short and cool. Granite of the Canadian Shield is interspersed with numerous lakes and swamps. The dominant natural vegetation is Boreal forest.

Nomadic hunters and gatherers, the ancestors of the Anishinabe and Cree populations that the Europeans encountered when they arrived, occupied the region for several thousand years after the last glaciation. The Europeans who first entered the region in the late 1600s were French fur traders and explorers. Permanent European settlement did not take place until the nineteenth century as the region's mineral and timber resources became more important. Silver was discovered in the

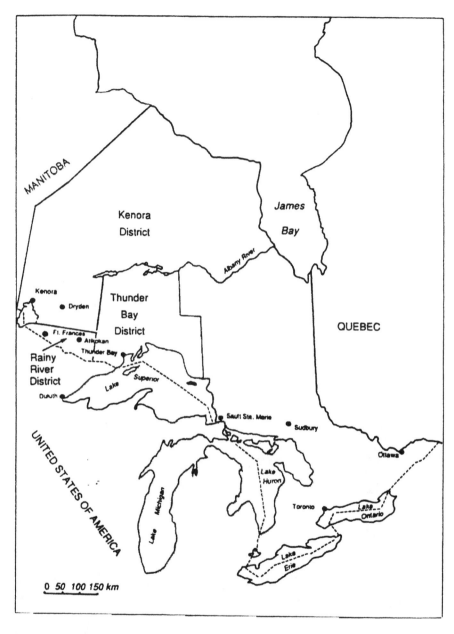

Figure 3.1 Ontario-Selected Urban Centres and Northwest Districts (*source:* Thomas Dunk).

1860s. The transcontinental railroad was completed in 1885 and a port was constructed at what is now the city of Thunder Bay for the transshipment of grain from the Canadian prairies. These developments stimulated European settlement in the region.

Northwestern Ontario became heavily dependent on industries involving the extraction, transportation, and minimal processing of natural resources. Typically, the ownership of these enterprises was, and continues to be, located outside the region in metropolitan centres in southern Ontario and Quebec or in the US. Prior to the spread of the welfare state in the post-Second World War era, the local population generally served as independent contractors and/or wage labourers for these industries. The male working-class culture that came to form the hegemonic masculinity of the region was shaped by the nature of work in these industries and by the pattern of reliance on external markets and investment capital. This working-class masculinity was a subordinate form of masculinity in the national society but in the region it was and, in many respects, remains dominant.

Unregulated labour relations typical of competitive capitalism characterized the first phase of modernization in the region. The work groups in the railroad and later highway construction camps, in the mines, logging camps, pulp and paper mills, and grain elevators were almost exclusively male. Much of this employment was physically demanding, dangerous, and required few formally recognized skills. It was also very seasonal and given to boom-and-bust cycles. Non-English-speaking workers who immigrated to the region from Quebec, or southern, eastern and northern Europe often filled the least secure and most dangerous occupations. Workers'

Figure 3.2 People's Co-operative (*source:* Thomas Dunk).

efforts to organize were vigorously resisted by employers (Bradwin 1972; Dunk 1996, 2003; Radforth 1987).

Working-class frustration was manifested in the development of a strong tradition of political radicalism that remained significant into the 1970s (Weller 1977). Socialist and communist tendencies marked working-class politics in the early part of the twentieth century. This was the product of a mixture of the moderate socialism of British trade unionism with the socialist and communist traditions brought from Finland and eastern Europe by immigrants from these parts of the globe. Greek and Italian immigrant workers also exhibited a propensity for what was at times violent job-related action (Morrison 1976).

In the years after the Second World War, the industrial base of the region remained rooted in resource extraction industries and transportation, although the expansion of education, health, and government services did somewhat diversify the occupational structure of the region. 'Fordist' labour relations were established in the larger enterprises. The railways, grain elevators, mines, and the pulp and paper industry were unionized. Increases in real wages and the creation of public pensions and public health care provided stability for working-class people in the region. Indeed, in some industries workers did very well, with wages rising to well above the national industrial average (Dunk 2002a). Relations between workers and employers could still be tense and strikes continue to be common but the radicalism of the labour movement in the region subsided under the complementary influences of economic growth, bureaucratic unionism, and employer and government sponsored suppression of communist union activists.[1]

Figure 3.3 Unionized industries (*source:* Thomas Dunk).

The meaning and significance of ethnic and race relations in the region also changed. With a reduction in rates of immigration from Europe in the 1950s and 1960s and the fact that the children of earlier waves of immigrants were educated in English, the ethnic divisions within the working class became less important. Where once ethnic networks and boundaries played significant roles in determining where one might work and live, and who one might marry, by the 1970s there was a growing sense of 'white' identity structured in relationship to the rapidly growing aboriginal population (Stymeist 1975; Dunk 2003). Residents of European ancestry have become what Stymeist refers to as 'name ethnics'; that is, they are conscious of their ethnic heritage and play it out in certain ritual forms during major holidays and important life events. However, the blatant ethnic division of labour which characterized the labour market between the later 1800s and the 1950s or 1960s, which all older residents still readily describe,[2] has faded considerably, although, racial divisions between the aboriginal population and whites are still strong. Indeed, male working-class culture in the region has a strong racialist element to it, as we discuss below.

By the late 1960s and early 1970s, as the long post-war boom experienced by the whole nation reached its end, a white, male working-class culture had become established. This culture was based upon relatively unskilled, physically demanding or uncomfortable, yet often, relatively well-paid employment in the resource extraction, transportation, and processing industries. These industries also provided significant employment for trades people such as mechanics, electricians, millwrights, plumbers, and pipefitters. The mostly white men who occupied these usually unionized jobs formed the local labour aristocracy. We turn now to a discussion of specific key elements of their culture.

Masculinity, class, and regional identity during the Fordist era

The pattern of capitalist development in northwestern Ontario was intimately linked to a particular form of working-class masculinity. By the 1970s, the regional male working-class culture was structured around a profound anti-intellectualism and celebration of commonsense, the informal group of male friends, and a white racial identity. All of these are linked with a sense of regional difference and grievance. None of these elements of local male working-class culture is unique to northwestern Ontario. However, they combine to form a particular structure of feeling in the region (Dunk 2003).

Anti-intellectualism and commonsense

Anti-intellectualism and the celebration of commonsense are key features of the regional, male working-class culture. These emphases are derived in part from the 'unskilled' nature of the work and the generally low educational levels of the male

36

working-class population in the region. However, it is not simply a reflection of low levels of formal education. Anti-intellectualism is a specific kind of response in the struggle over what constitutes legitimate and respected forms of thinking and modes of expression. It is part of a set of cultural practices and beliefs formed in opposition to the perceived characteristics of other cultural practices associated with those deemed by society for various reasons to have intellectual skills. The term anti-intellectualism here is to be understood as a way of thinking about the world and what really matters in it, a mode of approaching problems and issues that favours certain kinds of interpretations over others (see discussion in Dunk 2003).

This anti-intellectualism must be understood in terms of the division or gradient between mental and manual labour. Of course, all work involves both kinds of labour. Some kinds of work objectively do involve a larger mental component. However, the social process by which some kinds of work come to be accepted as involving a strong intellectual component or of being skilled is by no means simply a reflection of an objective measurement. There are a series of phenomenological, social, political, and historical factors that are involved (Attewell 1990; Dunk 1996). There is, in other words, a struggle over what counts as important forms of thought and knowledge. The division between mental and manual labour is reflected in terms of the differences in the occupational structure of the labour market in metropolitan southern Ontario and northwestern Ontario. In simple terms, there has always been far more white-collar employment in the south, while the north has been much more blue-collar. Working-class men in northwestern Ontario celebrate practical, down-to-earth skills and abilities over those related to, say, abstract thinking and theorizing. This corresponds to their sense of the people of the region as being hardworking and practical as opposed to the (admittedly stereotypical) image of people in metropolitan centres such as Toronto, Montreal, or New York as being impractical and frequently employed at jobs that do not require 'real' work. The metropolites may have a better formal education, but for the local working-class men this is no reason to respect them. To the contrary, their lack of 'commonsense' and practical skill is perceived to be one of the reasons why those from outside the region who have such influence over its affairs seem to frequently make decisions which harm local interests.

But while anti-intellectualism and the celebration of commonsense forms of thought are linked to a sense of class and regional difference, they also place limitations on the critical capacity of local working-class men. The emphasis on immediate sense experience and practical knowledge blunts the critical and more abstract thinking that is necessary to develop a deep understanding of the nature of the class and regional inequalities that have an impact on working-class lives in the region. This is seen in the ambivalent nature of other elements of working-class culture.

Male-bonding and the informal friendship group

Male working-class culture in the region continues to reproduce strong gender divisions. Occupations such as logging, mining, steveodoring, and railway and road construction that were typical of the early development of the region were physically demanding and sometimes mind-numbing. It required a grim toughness to survive them and the living conditions in the camps and early towns. Moreover, work sites were almost completely male in terms of their gender composition. These conditions, together with the cooperation between workers required by the labour processes involved, and the solidarity built through the struggles between local workers and their employers, meant that the male friendship group became an important source of identity and support for many working-class men. In the early phase of industrialization, these groups were also based on ethnic and linguistic commonalities.

The informal group of male friends is also an important focus of male working-class life outside of the workplace as well.[3] It was and continues to be based upon principles quite different than those of the wider capitalist society. The dynamics of interaction between members of the informal group illustrate an alternative set of cultural norms and practices. For example, in the wider society, exchange is based on the capitalist principles of competition and the maximization of one's own self interest. Within the informal men's group, however, exchange is based upon the principle of generalized reciprocity. This is evident in, for example, sporting activities. With regard to the purchase of drinks that inevitably is part of a typical outing, no one asks who has or has not contributed. There is no close record keeping or calculation. In the bar after a game, everyone takes turns buying rounds. If someone is short of money one evening someone else or several other people stand him drinks. This 'just happens'; one does not need to ask. Of course, participation in this system does contain an obligation. Everyone is expected to behave in the same manner. Anyone who is perceived to be too concerned with money, about who has paid for what, is the object of scorn. In this moral economy true friends are people who give to each other without asking, and this extends beyond the walls of the bar after a game, especially into the realm of do-it-yourself projects, and an entire underground non-market economy of exchanges of labour, knowledge (especially of practical skills that are highly valued), and goods.

The different values displayed and reproduced in the informal group are evident in other ways as well. There is an interesting correspondence between the competitive element of sport and the competitive features of a capitalist mode of production. Among the informal groups that we have studied in northwestern Ontario, many of the preferred sports have quite a different feel. The competitive nature of games such as baseball or hockey is de-emphasized. Of course, individuals do like to win but the competitiveness is tempered by the overriding principle of fun. There is an unstated but nonetheless clear recognition that competitiveness can create conflicts and anyone who takes the game too seriously is sneered at. The object of the game is to have fun and anything that might jeopardize this result is shunned (see the fuller

Figure 3.4 'The object of the game is to have fun' (*source:* Thomas Dunk).

account in Dunk 2003). The point of the game is to participate in a group activity, to have a laugh, to share a few drinks, and to cement friendships. It is a truly social ritual rather than an opportunity to exhibit one's individual prowess.

Although in the informal male group, one can see glimpses of an alternative culture based on principles such as solidarity, equality, and reciprocity (as opposed to the principles of competition, hierarchy, and market exchange that are dominant in contemporary capitalist culture), these groups also tightly maintain boundaries, especially against women and gay men. Much of this social closure takes place in bars, baseball fields, and curling and ice hockey arenas, in addition to the work sites, where women's roles are tightly controlled. Although women are slowly breaching the literal and symbolic mechanisms by which boundaries are maintained, this is still a world where women and gay men are excluded through the use of foul and sexist language and ethnic and gendered humour. The presumed role of women in this context is to provide support for the men, just as it still largely is within the domestic realm.

A specifically male solidarity is, thus, played out in the informal group. It reflects the masculine culture of typically all-male work places in its emphasis on equality and solidarity within the group and in its often sexist and homophobic elements that function to maintain group boundaries. It is in this sense an ambivalent social phenomenon. It is progressive insofar as it expresses working-class solidarity but it is also reactionary insofar as it is also the venue for the celebration of heterosexual masculinity in opposition to homosexual men and women.

The racialization of class and regional identity in male working-class culture

The ambivalent character of working-class masculinity in northwestern Ontario is also evident in the way in which regional, class and racial divisions are elided. As we have said, the ethnic divisions between immigrants with different European ancestries have become far less significant than they once were. At the same time, the sense of a white identity has become stronger. Whiteness is structured largely in relationship to the large and growing aboriginal population, the descendants of the first inhabitants of this part of the North American continent. White working-class masculinity is socially constructed in part around oppositions between white and Indian (white working-class men quite frequently consciously use this terminology rather than the currently more accepted terms such as aboriginal, First Nations, or Anishinabe and Cree).

In local, white, working-class culture, negative stereotypes of aboriginal people are common. On the one hand, this reflects a submerged but nonetheless present sense of racial superiority that is widespread among whites in Canada. In northwestern Ontario, it is also linked to a working-class critique of regional inequalities and the failure of the state to acknowledge the reality of class. From the perspective of white male workers, the aboriginal population receives preferential treatment from both the federal and provincial governments. The treaties signed in the nineteenth century, the Canadian Constitution, and a series of court decisions have affirmed the special status of aboriginal people in Canada. These provide aboriginal people with some avenues to redress economic inequalities that are not available to the rest of the population.

Despite their special constitutional status, overall aboriginal people comprise the least educated, least healthy, and poorest segment of the population. In socio-economic terms they are at the bottom of the social hierarchy in the region. Some of the local, white, working-class hostility towards aboriginal people reflects that combination of jealousy and fear that workers have of the 'welfare class'. Their own sense of status worth is constructed by strongly differentiating themselves from the social strata immediately below them. Moreover, commonsense dictates that one focuses on the easily observable and immediate. In practice this means that visible symptoms of deep problems are often taken as causes. In addition, local white workers resent the fact that governmental powers that rest in Ottawa and Toronto seem frequently to ignore the socio-economic plight of the largely working-class population of northwestern Ontario. The narrow·economic base, higher rates of unemployment, the lack of local control over the economy, and the lack of influence regional politicians seem to have on public policies that affect the region, are frequent subjects of complaint. All of these issues involve class and regional dimensions. Canada's official policy of multiculturalism and the unique constitutional status of aboriginal people in the country recognize ethnicity and race as potential sources of inequality and injustice. At the same time, they deny the reality of class inequalities.

The male, white working class in northwestern Ontario react against this dominant state ideology. They assert their own class identity and the importance of understanding class and regional inequalities. Unfortunately, they do so by emphasizing their perceived superiority over the most visible and poorest group in the region. The anti-intellectualism, which is itself an expression of a consciousness of class inequality and a reaction to it, blunts an interest in the more abstract, but nonetheless real, historical and structural factors that explain the poverty of the aboriginal population. The racialist tendencies in local, white, male working-class culture cannot be denied but nor can they be understood apart from a recognition of how they link to male workers' sense of class and regional inequality.

Post-Fordist realities and working-class culture

The anti-intellectualism which is expressed in local, male, working-class culture is a product of the split between mental and manual labour which is based on cultural practices and rituals that separate those with a certain kind of training, those who have successfully completed a set of rituals and who therefore have credentials, from those who do not. The division between mental and manual labour reflects real differences in educational levels and occupational opportunities between the northwestern and the southern, metropolitan zones of Ontario. Recent economic changes are likely to reinforce the regional differences because high technology, knowledge-based industries are clustered around larger metropolitan centres while resource hinterlands such as northwestern Ontario experience ongoing deindustrialization. The relatively lucrative, unionized jobs of the older resource extraction industries are declining in number. To the extent they are being replaced at all, it is with low wage, non-unionized jobs in telecommunications (telephone call centres for example) or the tourist industry. The other major source of employment in the region – federal, provincial, or municipal governments – has been cut back in the last two decades as governments downsize or contract out services.

The local working class is heavily affected by these changes. The male working-class culture that developed during the industrial era was based around male solidarity as expressed in the informal male group and a strong anti-intellectualism. This male working-class solidarity was also strongly sexist and racist. It has not equipped the local male workers to react in a progressive manner to more recent changes. The natural resources that once provided the region's economic base in some cases have been exhausted while in other situations changes in technology and markets have reduced demand for them. Moreover, social movements such as environmentalism, which are largely based in southern, metropolitan centres, have been able to bring about some changes in government policy the immediate economic impact of which have been negative in northwestern Ontario.[4] The explanation of these changes lies in global developments that require moving beyond immediate commonsense thinking to comprehend. The anti-intellectual features of regional working-class culture tend to blunt workers' interest in these more abstract historical processes (Dunk 1994, 1998, 2002a, 2002b). These developments also threaten to undermine the male solidarity that has been the bedrock of the regional masculine working-class culture for a century. Neo-liberal economic policies and neo-conservative social and political opinions which celebrate individualism and competition have, in the context of significant deindustrialization, divided workers into two camps, those who manage to survive and succeed in the new economy, and those who do not (Dunk 2002a). The working class in the region is being remade to fit this new world. Unless male workers are able to transcend the boundaries of gender and racial politics, they will continue to find themselves marginalized and divided, and ultimately powerless to resist the forces of change that are sweeping through northwestern Ontario as well as the rest of the world.

Conclusions

Working-class masculinity in northwestern Ontario thus illustrates the inextricable linkage between the forms and hierarchical arrangements of masculinity and social class that Connell (1995), Pyke (1996), and Wright (2001) have suggested. It also demonstrates the key role that spatial relations play in the never-ending processes of class and gender identity formation. Male working-class culture reflects the political and economic circumstances in which it developed, conditions that are intimately tied to the natural resources in this beautiful but environmentally harsh region. The emphasis on physical toughness, male solidarity, anti-intellectualism, commonsense, and an ethnic identity as whites reflect the structural realities workers confronted. These cultural elements remain as sediments that have their unique influence on male workers' responses to recent economic and political forces that are fundament-ally transforming the regional society. As these changes unfold, the cultural practices that mark hegemonic masculinity in the region and that differentiate social classes and regional identities will continue to evolve.

Notes

1 This is a complicated history and there is not space to discuss it adequately here. But see Radforth (1987) for a discussion about the suppression of communists among the loggers in the region.
2 The English and Scots had control of the best jobs on the railroads and grain elevators. Finns and French-Canadians were dominant in logging. Italians are said to have controlled construction of all kinds but particularly cement work and road and railroad construction. Ukrainians, Slovaks, and other eastern Europeans were employed on the docks as steve-dores, as navvies in rail and road construction, and in other occupations involving heavy manual labour.
3 Of course, the family is the center of life for many working-class men outside of the workplace. However, the male informal group competes for the attention of many male workers. This is a very common source of tension within marital relations.
4 Of course, in some cases these changes may have been necessary to the longer-term viabil-ity of some industries. Changes to logging regulations aimed at making better and more sustainable use of the forest possible are an example. However, the short-term impact of

these is often to increase costs to industry with the resulting decline in employment as the industry suffers economic difficulties. This is a sign of the economic and environmental squeeze that regions such as northwestern Ontario are in. There are no easy ways out of these dilemmas.

References

Attewell, P. (1990) 'What is skill?', *Work and Occupations* 17(4): 422–48.

Bradwin, E. (1972) *The Bunkhouse Man*, Toronto: University of Toronto Press.

Connell, R.W. (1995) *Masculinities*, Berkeley: University of California Press.

Dunk, T. (1994) 'Talking about trees: environment and society in forest workers' culture', *The Canadian Review of Sociology and Anthropology* 31(1): 14–34.

Dunk, T. (1996) 'Culture, skill, masculinity and whiteness: training and the politics of identity', in Dunk, T., McBride, S. and Nelsen, R.W. (eds) *The Training Trap: Ideology, Training and the Labour Market*, Halifax: Fernwood.

Dunk, T. (1998) 'Is it only forest fires that are natural? Boundaries of nature and culture in white working class culture', in Sandberg, E.L.A. and Sörlin, S. (eds) *Sustainability, the Challenge: People, Power and the Environment*, Montreal: Black Rose Books.

Dunk, T. (2002a) 'Remaking the working class: experience, class consciousness, and the industrial adjustment process', *American Ethnologist* 29(4): 878–900.

Dunk, T. (2002b) 'Hunting and the politics of identity in Ontario', *Capitalism, Nature, Socialism* 13(1): 36–66.

Dunk, T. (2003, 1st edn, 1991) *It's a Working Man's Town: Male Working-Class Culture in Northwestern Ontario*, second edition, Montreal: McGill-Queen's University Press.

Morrison, J. (1976) 'Ethnicity and violence: the lakehead freighthandlers before World War 1', in Kealey, G.S. and Warrian, P. (eds) *Essays in Canadian Working Class History*, Toronto: McClelland and Stewart.

Pyke, K. (1996) 'Class-based masculinities: the interdependence of gender, class and interpersonal power', *Gender and Society* 10(5): 527–49.

Radforth, I. (1987) *Bushworkers and Bosses: Logging in Northern Ontario 1900–1980*, Toronto: University of Toronto Press.

Stymeist, D. (1975) *Ethnics and Indians: Social Relations in a Northwestern Ontario Town*, Toronto: Peter Martin Associates Limited.

Weller, G. (1977) 'Hinterland politics: the case of Northwestern Ontario', *Canadian Journal of Political Science* 10(4): 727–54.

Wright, E.O. (2001) 'A conceptual menu for studying the interconnections of class and gender', in Baxter, J. and Western, M. (eds) *Reconfigurations of Class and Gender*, Stanford: Stanford University Press.

Further reading

Fawcett, B. (2003) *Virtual Clearcut, or, the Way Things Are in My Hometown*, Toronto: Thomas Allen. This is a book about northern British Columbia but the nature of the community and the economic and social issues are very similar to the northern Ontario situation.

Livingstone, D.W. and Mangan, J.M. (eds) (1996) *Recast Dreams: Class and Gender Consciousness in Steeltown*. Toronto: Garamond Press. A collection of sociological essays on the linkages between working-class culture and masculinity among steelworkers in Hamilton, Canada's largest steel-making centre.

Ondaatje, M. (1987) *In the Skin of a Lion*. Toronto: McClelland and Steward. This is the prequel to his more famous novel, *The English Patient*. It follows an Eastern European immigrant from Toronto up into the bush camps in northern Ontario. A beautifully written novelistic treatment of life among the itinerant male immigrant workers.

Radforth, I. (1987) *Bosses and Bushworkers: Logging in Northern Ontario 1900–1980*, Toronto: University of Toronto Press. This is the definitive labour history of logging in the region. Very good on the struggles between red and white Finns in northern Ontario.

4

MASCULINITY AND RURALITY

Paul Cloke

Summary

In this chapter, I will pursue a number of significant traces in the interconnections between rurality and masculinity. By way of introduction, I will contextualize the development of rural studies in terms both of an emergent emphasis on patriarchy and gender (though relatively underdeveloped on masculinity per se), and in terms of the aggregate effects of masculine bias among most rural researchers and theorists. The chapter will then discuss three strands of research on masculinity: the representational and experiential masculinities of agricultural societies; the masculinities involved in rural problematics, especially poverty and homelessness; and the 'alternative' masculinities of gay sexuality in the countryside.

- Recognizing masculinity in rural settings
- Relating rurality and masculinity
- Different traces of masculinity in/of the rural
- Multiple rurals/multiple masculinities: the importance of identity.

Introduction: recognizing masculinity in rural settings

> The distinctive character of studies of masculinity lies less in their theoretical originality than in their ability to reconfigure, and otherwise shed light on, ideologies and practices that have traditionally been considered unrelated to matters of gender.
>
> (Bonnet 1999: 160)

Let's start this chapter close to home. As a male academic who over the last 25 years has been involved in researching and writing about rural areas, I am acutely aware that the academic disciplines contributing to the loose collectivity of rural studies have been dominated by men. Whilst fully acknowledging the outstanding contribution of key women researchers and commentators to rural studies, I believe (following Rose 1993, 1999) that women have been marginalized as producers of knowledge about rurality and rural areas, with considerable consequences not only for what counts as legitimate knowledge, but also for the unspoken focus on particular spaces, places and landscapes of the rural which have particular relevance to men. That is, the preponderance of rural studies has occupied a masculinist subject position and has exhibited 'malestream' thinking. The first setting, then, in which there is a need to recognize and reconfigure the ideologies and practices of masculinity is that of the rural knowledge industry itself.

As recently as 1986 Jo Little was having to point out the dearth of published work in rural geography on both gender issues generally, and more specifically on the lives of rural women. In so doing, she argued for a new emphasis on gender difference and inequality in rural societies and began to outline how feminist perspectives could revolutionize theoretical and conceptual debates in rural studies. Her recent book on *Gender and Rural Geography* (Little 2002) gives testimony to the very significant impacts which have been made by geographies of gender since that time. However, as she concludes, 'a lot of ground has been covered . . . but there are still major gaps' (p. 3). Although at least some of the old misogynist attitudes have disappeared (or are in hiding) and increasing numbers of (particularly younger) researchers have been willing to espouse and embrace gender as a key foundation of rural knowledge, there remains a widespread sense of sometimes grudging 'permission' by men for engagement with gendered studies of the rural, rather than an 'ownership' of such studies. It is almost as if by accepting the legitimacy of studying rural women, underlying gender relations in academic and rural worlds need not be unduly disturbed.

This is why the emergence of understandings of masculinity is so important. Studying masculinity is not a pathway to reinforcing the hegemonic focus on men to the detriment of gendered understandings of women. Far from it! A focus on the lives, experiences, assumptions and unspoken controlling influences of men will inevitably be relational, inducing a clearer grasp of how gender relations are managed, negotiated and performed, and of how male and female experiences and expectations interconnect. As Bonnet suggests, studying masculinity will also shed light on aspects both of studying rural areas, and more particularly of rural areas themselves, which previously had not been identified as issues of gender. Masculinist mastery of knowledge and power over discursive coding of places and practices can be demystified and disrupted by the corporealization of the performance of that knowledge and those spaces; 'once embodied as, for example, white, male, straight or able-bodied, the intrepid explorer and his voice of reason lose their transparent cloak of neutrality . . .' (Rose 1999: 249).

More generally, there has been a surge of interest in masculinity over recent years (see Berger *et al.* 1995; Brod and Kaufman 1994; Clare 2000; Connell 1987, 2000; Guggenbuhl 1997; Hearn 1998; Kimmel 1996; Mac an Ghaill 1996; Rutherford 1997; Sweetman 1997; Whitehead 2002; Whitehead and Barrett 2001). In comparison there has been a relatively slow adoption of masculinist foci and approaches in rural studies, with a landmark being a special issue of the journal *Rural Sociology* (Bell and Campbell 2000). In this chapter I want to address two sets of questions in the light of some of the published studies of rural masculinity. First, in what different ways do rurality and masculinity interconnect? Clearly, there are many different ruralities (Cloke 1999), and ways of envisioning rurality (Cloke 2003), just as there are no homogeneous masculinities. Equally masculinity is part cultural construction and part material reality, and will intersect with geographical space and imagined space in different ways (see Cloke and Little 1997). Second, what are the different traces which emerge in/of the rural according to varying conceptualizations of masculinities? Only by addressing these questions will an appropriate sense of multiple rurals and multiple masculinities emerge.

Relating rurality and masculinity

In order to explore different relations between rurality and masculinity, I want to use by way of illustration two excellent studies of masculinity in rural New Zealand. The first by Robin Law, a geographer from Otago (now sadly no longer with us), examined how an advertising campaign for a local South Island beer – Speight's – constructed a form of rural masculinity which became know as 'Southern Man'. Acknowledging previous work on the playful use of place-indicators and gender identities (see, for example, Jackson and Taylor's 1996 study of advertisements for Boddington's beer in the UK) she sought to draw out how particular constructions of masculinity become socially dominant and are legitimized as being somehow a 'natural' gendered order. She demonstrated how conventional notions of 'real men' in 'unspoiled nature' were offset in the campaign by a self-mocking humour which imparted a particular personality of masculinity:

> In the first and most successful of the three TV commercials, a young man and an older man are resting at dusk by a lake. Both speak with the distinctive Southland burr, using local phrases ('She's a hard road . . .') and in a laconic style punctuated with many silences. The younger one reveals that he's been seeing a city girl who wants him to move up to Auckland. She can offer a range of material comforts . . . but she doesn't drink Speights beer and that clinches it. The younger man's decision is affirmed by the older man's encouraging words: 'Good on yer, mate'.
>
> (Law 1997: 24–5)

At one level, the advertisements portray the 'mateship' of single Pakeha (white) heterosexual males in the outdoor environment of hard physical labour and intergenerational mentoring. As Law reveals, however, several place-related dimensions are built into the imaginative construction of 'Southern Man'. First, a number of signifiers – not least landscape, land use, social conventions, employment, clothing and strategic presence of farm-related animals – ensure that the advertisements are located in 'the rural'. In this way they enrol rural cultural traditions which speak of the authenticity of nature, and the masculinity of New Zealand rurality (Berg and Kearns 1996). Second, there are clear indications of particular regional rurality, denoting the High Country of Central Otago where the purity and hardness of the country can be promoted as model characteristics for wider New Zealand masculinity. Third, the advertisements portray a timeless place: 'Among the images of "unspoiled" nature, stripped-down male bodies, companion dogs and horses, there are no disturbing modern gadgets such as farm bikes, two-way radios, or binoculars; not even a plastic bag' (Law 1997: 25).

Timelessness evokes nostalgia for a more simple way of life which is threatened by various forms of progress. Rural life is becoming increasingly marginalized economically, and masculinities of bravery, audacity and strength of character are both anachronistic yet appealing in terms of their authentic values.

The Southern Man campaign quickly took root in a material culture of billboards, posters, T-shirts and other commodified paraphernalia, which illustrated further both the character of Southern Masculinity and the relations between masculinity and urban/cultural 'others'. Figure 4.1. details the tongue-in-cheek text of a Speight's T-shirt displaying a 'Southern Man Identification Chart' the basis of which is that 'it's time we mentioned and reminded ourselves of the types of men (and their dogs) that make this country great'. Law suggests that the images and phrases of the campaign have become part of the cultural repertoire of New Zealanders and are 'used and reworked to support, satirize and comment on the validity of this construction of regional identity and masculinity' (p. 26).

The second illustrative study is by Hugh Campbell, a social anthropologist, also based at Otago University in South Island, New Zealand. Drawing on work with other colleagues (Campbell and Phillips 1995; Fairweather and Campbell 1990) he analyses male pub drinking practices, not as a nostalgic memorial to a simpler life, but as a site in which the power and legitimacy of masculinity are cemented in rural community life. His study (Campbell 2000) focuses on a small South Island town, where a core group of 150 to 200 men working in farming or farm-related industries regularly congregate for after-work drinking in the public bars of two local hotels. Other social groups – ski industry workers and tourists, white-collar professionals, and women – were excluded from these gatherings. Indeed Campbell notes that if women were accompanying the man back home from work, they would either drink separately in the lounge bar, or spend the time shopping or even sitting in the car with the kids. Thus, the drinking routine takes on an important emphasis:

> The significant feature of this group is that the men who gathered in the pubs also tended to be the most influential members of sports clubs, charities, the local council and (non-tourist related) business. The after-work drinking session was not an isolated enclave of male activity, but the centre of a network of association among men in the community.
>
> (Campbell 2000: 569)

In other words drinking beer in the pub after work constitutes a crucial performance of public masculinity at which men's power in the community is defended and

Figure 4.1 Southern Man Identification Chart (*source:* Speight's www.speights.co.nz/
images/sthn_man.swf)).

continually legitimized. Through unwritten codes which establish the necessary disciplines of drinking and competitive banter (Campbell calls this 'conversational cockfighting') the invisible framework is erected by which a 'real man' may be defined, and how a real man behaves may be discerned. This performative construction of masculinity becomes 'part of the unquestioned fabric of social interaction' (p. 578) and demonstrates more ideological relations between rurality and masculinity:

> The male drinkers' notion of locality, however, contains all the attributes of what might be regarded ideologically as rural. Locality in their conception, is characterized by long-term residence and by farming or primary industry-related employment; is organized predominantly around kinship rather than clan groups; is virtuous and morally superior to city life; and represents a more 'natural' social order for human life. Therefore the pub(lic) performance of hegemonic masculinity not only is mediated through the structural characteristics of a specifically rural community, it also mobilizes notions of locality/rurality that reinforce the embeddedness, and thus the 'naturalness' of this particular version of masculinity.
>
> (p. 579)

These studies by Robin Law and Hugh Campbell are not only examples of seminal research in their own right, but together they illustrate some of the fundamental relations between rurality and masculinity, more generally. These relations have been categorized in terms of the difference between the 'masculine rural' and the 'rural masculine' (Campbell and Bell 2000). The masculine (in the) rural, refers to the ways in which masculinity is socially constructed in different kinds of rural spaces. The rural (in the) masculine refers to the manner in which particular characteristics or significations of rurality help to construct ideas of masculinity. Campbell's male after-work drinkers offer a clear illustration of the masculine rural, constructing community power and performative norms in terms of a dominant group of men, and in turn promoting the invisibility of others, notably women. Law's portrayal of 'Southern Man' reflects the rural masculine, demonstrating how particular discourses of rurality become incorporated into popular cultural interpretations of masculinity. Yet (as Campbell and Bell acknowledge) these categories should not be regarded as mutually exclusive. The cultural constructions of rurality inherent in the Southern Man campaign were translated into material merchandise, conversational habits and even behavioural replication which all mark a cross-over from the rural masculine into the masculine rural. Similarly, the masculinist performances of the after-work drinkers were characterized by particular ideas about ideal-type rurality, especially in terms of the virtues and moral superiorities of rural life in comparison to its city other. Thus the masculine rural crosses over into the rural masculine.

Sometimes it is all too easy to pigeon-hole the issues of masculinity in rural areas either in terms of cultural environments which are somehow 'out there' or in terms of individual activity which is 'right here'. My argument would be that such categorizations are effectively indivisible. As Whitehead (2002) suggests:

> Cultural environments are not 'out there', somehow existing external to the individual, but are (in)formed by individual subjects, though not necessarily in cognitive fashion. In the very moment that individual action impacts on the social, so a cultural environment is created – local and temporary as it might be.
>
> (p. 13)

It is in these discursive moments that the complex interrelations between rurality and masculinity are most crucially displayed.

Different traces of masculinity in/of the rural

Studies of rural masculinity have moved quickly through a number of conceptual stages, beginning with a critique of apparent essentialisms about both masculinity and rurality, then dwelling considerably in the territory of social construction in order to understand masculinity in terms of a process of socialization into particular roles and expectations, and more recently beginning to explore the discursive moments when individual actions impact on social and cultural environments. To talk of essential characteristics of rural masculinities flies in the face of the widely acknowledged variations pertaining both to rurality and masculinity which are plural and multiple, differing over space, time and context. Nevertheless, early understanding of the importance of these masculinities in their rural contexts drew heavily on sociological notions that 'men have qualitatively different characters to women – that men and women each have their own separate and discrete spheres of social experience and power' (Bonnet 1999: 160).

In rural studies, these significant differences between men and women were initially exposed in terms of how uneven patriarchal power relations permeated not only the agricultural sector, but also wider aspects of rural community life. In essence, such unevenness was portrayed in terms of the contrast between the authority and controlling influence of men both in work and at home, and the expectation that women would accept a subordinate role which serviced and upheld the traditional processes, practices and expectations of life in a rural setting. Masculine power and feminine subordination, then, represented the unseen orthodoxies of social relations in the countryside, and these relations spawned ideologies and practices which for some were perceived as a naturalized order of things. Thus key spaces and practices were understood as somehow naturally men's spaces and practices (see Dempsey 1990, 1992), as illustrated in our previous discussion of after-hours drinking in a New Zealand rural town. Moreover, patriarchy was seen to extend to a male dominance over rural environments more generally. As Little and

Panelli (2003) have suggested: 'Associating femininity with nature and rural land-scapes has resulted in the perpetuation of the male gaze and the notions of male hus-bandry, dominance and control of rural and wilderness settings' (p. 284).

In this way, wild rural spaces become seen as male spaces, places for masculinized adventure which conflate a cultural mastery of nature with particular practices of exploration and adventure in which boys and men can perform brave feats to confirm and reconfirm their masculinity (see Cloke and Perkins 1998 on adventure tourism; Jones 1999 on gendered childhood adventure; Morin *et al.* 2001 on mas-culinity and mountains; and Phillips 1995 on Victorian masculine adventure). Rural areas even offer the ideal place in which to reclaim and renaturalize unreconstructed forms of masculinity (see Bonnet 1996 on the mythopoetic men's movement; and Kimmel and Ferber 2000 on right-wing militia in rural America).

Any inherent essentialisms in these men's spaces, practices and identities have quickly been challenged on the grounds that these malenesses are not the result of essentially masculine biologies or psychologies which somehow induce a kind of organic sex role which dictates a masculine power position. Connell's (1987, 1995, 1998, 2000) groundbreaking contribution to this challenge was to conceptualize the construction of masculinity as a social phenomenon, and to argue that the hegemonic position of masculinity will vary according to different contexts and relations of con-struction. 'Hegemonic masculinity is not a fixed character type, always and every-where the same. It is, rather, the masculinity that occupies the hegemonic position in a given pattern of gender relations' (Connell 1995: 74).

While retaining the focus on male power, hegemonic masculinity signals a multi-plicity of different and often contested male practices. Some men will exhibit com-pliant, subordinate or marginalized expressions of masculinity which may go against the grain of hegemony. Equally, women may challenge and resist hegemonic mas-culinity even though located 'within a balance of forces, that is a state of play' (Connell 1987: 184).

This notion of hegemonic masculinity helps make sense of the obvious non-essen-tialisms of masculinities in rural settings. Clearly not all men thrive in rural spaces. For example, Ni Laoire's (2001) study of young men who 'stay behind' in rural Ireland highlights a marked disparity between the persistence of masculine hege-mony, and the often tragic marginalization of certain groups of rural men for whom economic restructuring and challenges to traditional male orders lead to precarious personal positions from which severe stress and even suicide can result. Smith's (1994) study of farmers' responses to agricultural restructuring in New Zealand during the 1980s and 1990s similarly identifies men whose supposed hegemonic position is turned upside down by the social and psychological, as well as economic, consequences of 'walking off the land'. Other obvious non-essentialisms are evident in the roles played by some women in rural life. For example, the exploration of how 'farm wives' performed vital contributions to farm businesses (Gasson 1992; Sachs 1983; Whatmore 1990) has challenged orthodox understandings of gender relations in agriculture, and has given particular credence to studies which identified a key role performed by women in environmental activism and wider protest in rural areas (see, for example Liepins 1998; Mackenzie 1992).

The challenge to essentialism has been linked conceptually to a growing appreciation of how masculinity and rurality are intertwined by social construction. Thus overall constructions of rural life – such as that of the rural idyll (see Cloke and Little 1997; Cloke and Milbourne 1992; Little and Austin 1996; Mingay 1989) – serve to emphasize the inherently positive qualities of rural living while hiding the equally inherent expectations of traditional gender relations and culturally gendered practices which also underpin and sustain the idyll-ized way of life (Hughes 1997a, 1997b). Representations of rural culture, then, produce imagined geographies of rural areas which resonate with seemingly natural orders of things in which the invisibly powerful hand of masculinity renders invisible or marginalized those non-hegemonic aspects of rural gender relations.

The recent surge of interest in rural masculinities has explored some of the key social constructions of relations between rurality and masculinity. Two illustrative examples will be used here. First, considerable attention has, inevitably, been devoted to masculinism in agriculture, as in Liepins' (2000) study of media representations of agriculture in Australasia. She concludes:

> The dominant forms of masculine representation in Australian and new Zealand agriculture, at both farm and industry levels, show that notions of strength and battle are crucial elements in these media and industry discourses. Other aspects, including attention to outdoor labour, decisiveness, and key industry positions, are also relevant to these masculinities. Together these meanings are attributed to the subjectivities of 'farmers' and 'farm leaders' in media and farming organizations. Apparent 'truths' are built up cumulatively around these positions, which then normalize the 'truths' in association with 'tough men' and 'powerful leaders'.
>
> (p. 617)

In this way, seemingly universal truths are constructed about the maleness of agricultural space and practices, and key aspects of masculine identity – toughness, strength, ability to battle and so on – are intertwined with the very nature of rural land and outside work. These social constructions are reinforced by symbolic mediations of rural/agricultural life. So, for example, Brandth (1995) and Saugeres (2002) have shown how the tractor serves as a symbol of powerful masculinity in agriculture, in opposition to both femininity and nature. The appropriation of agricultural technology creates a symbolic masculine space in which women become excluded from the supposed centre of agricultural production. Discourses linking men with physical strength and aptitude for technology serve to legitimize unequal gender relations in agriculture, and construct a common sense world of rural practice which implies that women are not, for example, meant to be driving tractors. Interestingly, Brandth's (1995) study notes that as farm technology becomes more computerized and comfortable, so the picture of hegemonic agricultural masculinity is also being reconstructed into a more 'business-like' (p. 132) form. Similar examples of how masculinity is incorporated into representations of rural land can be found in studies of the forestry industry (see, for example, Brandth and Haugen 1998, 2000).

The second example of socially constructed relations between rurality and masculinity can be found in Woodward's (1998, 1999, 2000) study of military masculinities as a form or rural masculinity. Here, too, representations of rural masculinity identify the physical strength and power of 'real men' who are trained to master the outdoor environments of rurality. Woodward's focus is on 'the warrior hero':

> The warrior hero is physically fit and powerful. He is mentally strong and unemotional. He is capable of both solitary, individual pursuit of his goals and self-denying contribution towards the work of the team. He's also a bit of a hero with a knack for picking up girls and is resolutely heterosexual. He is brave, adventurous and prepared to take risks. Crucially he possesses the abilities to conquer hostile environments to cross unfamiliar terrain and to lay claim to dangerous ground.
>
> (Woodward 2000: 643–4)

In this instance, masculinities of tenacity, strength and decisiveness are co-constructed with rugged outdoor rural environments. As Woodward emphasizes, rural locations are specifically selected for training so as to construct and mould a specific kind of soldier. Thus 'the rural constructed in military training is matched to the masculinities exemplified in the strong, brave, hard warrior hero' (p. 652).

These examples convey naturalized and hegemonic forms of masculinity which are made believable by the ruggedness of the rural outdoors. They legitimize ideas about certain kinds of men being 'in place' in the countryside, and about the countryside being 'the place' for these masculinist practices and identities. Stereotypes emerge about how to be a man in the countryside, and these in turn become coded in common-sense norms for rural behaviour and attitudes. However, studies of rural masculinity have also sought to decentre these hegemonic stereotypes, often by focusing on rather different senses of desire and embodiment which link with alternative constructions and naturalization of men and masculinity. I will briefly mention three such unsettling moments here.

First, in a context where many rural areas are no longer dominated economically by primary land-using industries or landscapes of rugged wildness, there are now many places where social recomposition has brought significant numbers of in-migrant households into the rural domain. A series of studies of 'new' middle-class presences in the countryside (see Cloke and Thrift 1987, 1990; Cloke et al. 1995, 1998; Phillips 1998a, 1998b, 1999, 2002) have delineated a number of different identities adopted by in-migrants and pursued as 'lifestyle discourses' in the 'cultural texture' of village life. These include the village gentry lifestyle, the move in and join in lifestyle, the move in for self and show lifestyle and the village regulator lifestyle. If ideas of gentry and regulation perhaps reflect a reconstitution of previous roles, the purposive moving in either to join with local community, or particularly for a more detached form of displayed conspicuous consumption suggest the forging of new or renewed identities in rural places. As Cloke and Thrift (1987) discuss, traditional gender relations can be disturbed, for example, by the influx of dual income households, or by any sense in which in-migrants are unwilling to 'join in' with village life and values. This is not to presuppose that these disturbances to traditional relations will necessarily undermine patriarchal power in the countryside – what is happening here may be interpreted merely as a shift in localized hegemonic masculinity. However, these new forms of social relations often represent a professionalized, consumption-driven masculinity which is a far cry from that of the tough farmer or the warrior hero. Indeed such differences may themselves be revealed in local contests between newcomer populations and those engaged in extensive industrialized (and perhaps environmentally questionable) use of rural land.

Second, alternative desires and embodiments of masculinity are evident in how rurality can be used to naturalize gay masculinity (Bell 2000a, 2000b, 2003; Bell and Valentine 1995). Here the construction of countryside can be seen as sexualized or eroticized in a number of different ways in the gay imaginary. According to Bell (2000a) these constructions are:

> set apart in that they also open up a space for certain forms of same-sex activity whether this is described as natural 'manly love', hillbilly priapism, or rustic sodomy. Here is a rare and particular set of cultural circumstances, male same-sex activity and desire can be conceptualized as 'closer to nature' rather than as 'against nature'. As a result, homosexuality is essentialized somewhat problematically with this construction.
>
> (p. 559)

Recognition of these sexualized constructions of rural spaces and practices again serves to decentre and unsettle stereotypical accounts of rural masculinity. While hegemonic masculinities in many rural areas may well be performed in a closely guarded heterosexual terrain, Bell's work provides us with other versions of how rural masculinity can be constructed. The potential impact of unsettling rural male stereotypes is such that it has frequently been picked up by the advertising media as a source of shock impact. Figure 4.2. shows an advertisement for Malvern mineral water, which – using the banner 'Malvern. Not quite middle England' – displays an image which represents a decentring of the rural. What might otherwise be a stereotypical rural scene – two older men, dressed in tweed jackets and cloth caps sitting at the table of a traditional English pub – is symbolically reconstructed by the holding of hands and the strategic positioning of gay icons around the scene. This particular view of a typical masculine place seeks to subvert the construction of hegemonic masculinity, but perhaps merely succeeds in reinforcing that hegemony by its all-too-knowing references to sexual otherness.

Figure 4.2 Malvern 'Not Quite Middle England' (*source: Guardian Weekend*, 12 July 2003, p. 39. 'Malvern' is a registered trademark of Atlantic Industries).

Third, further examples of alternative embodied desires are to be found in accounts of the lives of marginalized men in rural areas. In studies of rural poverty (see Cloke *et al.* 1995a, 1995b, 1995c, 1997) and rural homelessness (Cloke *et al.* 2000, 2001, 2002), for example, there are strong indications of different configurations of rurality and masculinity. Although circumstances of social marginalization by no means nullify the powerful patriarchal position of men within households, their powerlessness in relation to economic capacity and to the institutions and gatherings of civil society stands as stark contrast to the spaces and practices of hegemonic rural masculinity. Even here, however, mainstream constructions of men and countryside cannot totally be evaded. Consider the following account of a young man sleeping rough in the countryside:

> I was going all over the place. Um, I'd walk round like 10, 15 miles a day in the countryside, just carry on walking ... Um, if I seen a rabbit I'd shoot it and um, find somewhere that was nice and isolated and then start cooking it. And then, once I'd eaten it, stay around for 5–10 minutes, and then, um, destroy the fire, make sure it looked like no-one had had a fire and then disappear.
>
> (quoted in Cloke *et al.*, 2002: 65)

Despite his relatively powerless position, this homeless man's relations with rurality conform somewhat uncomfortably with wider ideas around strength, craftiness, and mastery over nature. The principle decentring factor here is that his homelessness signifies an out-of-placeness in the countryside, which belies these other conformities. More generally, socially marginalized men will often escape supposedly idyllized constructions of their rurality.

Multiple rurals/multiple masculinities: the importance of identity

It is important to reiterate that not all rurality and masculinity can be encapsulated in studies of New Zealand beer-drinking, Scandinavian farming, British soldiering and the like. Hegemonic masculinity will clearly vary according to geographical and temporal contexts as well as in terms of particular patterns of gender relations. Although cultural constructions of rurality sometimes travel long distances via global media, social constructions of rural masculinity will be sensitive to more localized cultural prompts, and any attempts to universalize such constructions should be handled with care. However, in concluding, I want to move beyond social constructions for other reasons as well, returning to my previous argument about the effective indivisibility of the masculine rural and the rural masculine.

Masculinities will differ over rural spaces, times and contexts, and will be intertwined with other powerful social categories such as class, ethnicity, age and sexuality. They are to be found in mediated social constructions, but they are by no means dislocated from the social web. Rather they are implicated in the everyday lives, practices and spaces of men. As Whitehead (2002) has cogently argued:

> The entry of the subject into the social marks the point at which the prior codes of sex and gender signification 'go to work'. It is through the inculcation of multiple discourses across subject positions that the subject becomes

aligned with a prior politicized category: male/man. At this point the subject can be identified as a masculine subject. That is, through the imminent search for existence and being (male/man), the subject engages with and works on the historically and culturally mediated codes of masculinity that prevail around it. As these codes are already placed at the disposal of the subject, they offer a ready means of identity signification.

(p. 216)

Whitehead labels this process 'identity work' (p. 216). The search for being and becoming male or man comes via this identity work. I want to suggest that further understandings of the interrelations between rurality and masculinity will be likely if we take seriously this notion of identity work, for it is at this moment that social constructions and individual being/becoming seem to merge. Whereas the individual subject will bring particular senses of embodiment and desire to their identity work, that work is already largely prefigured by constructions of gender in which the individual is immersed because of their exposure to social and cultural circulation. The moment of identity work is therefore discursive, and this leaves open the political possibility of resistance and alternative practice. In this way, then, our further interrogation of dominant masculinities in rural areas can be a significant political contribution to the task of revising currently hegemonic discourses of gendered identity.

References

Bell, D. (2000a) 'Farm boys and wild men: rurality, masculinity and homosexuality', *Rural Sociology* 65: 547–61.

Bell, D. (2000b) 'Eroticizing the rural', in Phillips, R., Watt, D. and Shuttleton, D. (eds) *De-centring Sexualities: Politics and Representations Beyond the Metropolis*, London: Routledge.

Bell, D. (2003) 'Homosexuals in the heartland: male same-sex desire in the rural United States', in Cloke, P. (ed.) *Country Visions*, Harlow: Prentice Hall.

Bell, D. and Valentine, G. (1995) 'Queer country: rural lesbian and gay lives', *Journal of Rural Studies* 11: 113–22.

Bell, D. and Campbell, H. (eds) (2000) 'Special issue: rural masculinities', *Rural Sociology* 65: 532–658.

Berg, L. and Kearns, R. (1996) 'Naming as norming: "race", gender and the identity politics of naming places in Aotearoa/New Zealand', *Environment and Planning D: Society and Space* 14: 99–122.

Berger, M., Wallis, B. and Watson, S. (eds) (1995) *Constructing Masculinity*, New York: Routledge.

Bonnet, A. (1996) 'The new primitives: identity, landscape and cultural appropriation in the mythopoetic men's movement', *Antipode* 28: 273–91.

Bonnet, A. (1999) 'Masculinity/masculinities/masculinism', in McDowell, L. and Sharp, J. (eds) *A Feminist Glossary of Human Geography*, London: Arnold.

Brandth, B. (1995) 'Rural masculinity in transition: gender images in tractor advertisements', *Journal of Rural Studies* 11: 123–33.

Brandth, B. and Haugen, M. (1998) 'Breaking into a masculine discourse: women and farm forestry', *Sociologia Ruralis* 38: 427–42.

Brandth, B. and Haugen, M. (2000) 'From lumberjack to business manager: masculinity in the Norwegian forestry press', *Journal of Rural Studies* 16: 343–55.

Brod, H. and Kaufman, M. (eds) (1994) *Theorizing Masculinities*, London: Sage Publications.

Campbell, H. (2000) 'The glass phallus: pub(lic) masculinity and drinking in rural New Zealand', *Rural Sociology* 65: 562–81.

Campbell, H. and Bell, M. (2000) 'The question of rural masculinities', *Rural Sociology* 65: 532–46.

Campbell, H. and Phillips, E. (1995) 'Masculine hegemony in rural leisure sites in Australia and New Zealand', in Share, P. (ed.) *Communications and Culture in Rural Areas*, Wagga Wagga: Centre for Rural Social Research.

Clare, A. (2000) *On Men: Masculinity in Crisis*, London: Chato and Windus.

Cloke, P. (1999) 'The country', in Cloke, P., Crang, P. and Goodwin, M. (eds) *Introducing Human Geographies*, London: Arnold.

Cloke, P. (ed.) (2003) *Country Visions*, Harlow: Prentice Hall.

Cloke, P. and Thrift, N. (1987) 'Intra-class conflict in rural areas', *Journal of Rural Studies* 3: 321–34.

Cloke, P. and Thrift, N. (1990) 'Class and change in rural Britain', in Marsden, T., Lowe, P. and Whatmore, S. (eds) *Rural Restructuring*, London: David Fulton.

Cloke, P. and Milbourne, P. (1992) 'Deprivation and lifestyles in rural Wales: II Rurality and the cultural dimension', *Journal of Rural Studies* 8: 359–71.

Cloke, P., Phillips, M. and Thrift, N (1995) 'The new middle classes and the social constructs of rural living', in Butler, T. and Savage, M. (eds) *Social Change and the Middle Classes*, London: UCL Press.

Cloke, P., Goodwin, M., Milbourne, P. and Thomas, C. (1995a) 'Deprivation, poverty and marginalisation in rural lifestyles in England and Wales', *Journal of Rural Studies* 11: 251–66.

Cloke, P., Goodwin, M. and Milbourne, P. (1995b) 'There's so many strangers in the village now: marginalisation and change in 1990's Welsh rural lifestyles', *Contemporary Wales* 8: 47–74.

Cloke, P., Milbourne, P. and Thomas, C. (1995c) 'Poverty in the countryside: out of sight and out of mind', in Philo, C. (ed.) *Off the Map: A Social Geography of Poverty*, London: Child Poverty Action Group.

Cloke, P., Milbourne, P. and Thomas, C. (1997) 'Living lives in different ways? Deprivation, marginalisation and changing lifestyles in rural England', *Transactions of the Institute of British Geographers* 22: 210–30.

Cloke, P. and Little, J. (eds) (1997) *Contested Countryside Cultures*, London: Routledge.

Cloke, P. and Perkins, H. (1998) 'Cracking the canyon with the Awesome Foursome: representatives of adventure tourism in New Zealand', *Environment and Planning D: Society and Space* 16: 185–218.

Cloke, P., Phillips, M. and Thrift, N. (1998) 'Class, colonisation and lifestyle strategies in Gower', in Boyle, P. and Halfacree, K. (eds) *Migration to Rural Areas*, London: Wiley.

Cloke, P., Milbourne, P. and Widdowfield, R. (2000) 'Homelessness and rurality: "out-of-place" in purified space?', *Environment and Planning D: Society and Space* 18: 715–35.

Cloke, P., Milbourne, P. and Widdowfield, R. (2001) 'Interconnecting housing, homelessness and rurality', *Journal of Rural Studies* 17: 99–111.

Cloke, P., Milbourne, P. and Widdowfield, R. (2002) *Rural Homelessness: Issues, Experiences and Policy Responses*, Bristol: Policy Press.

Connell, R. (1987) *Gender and Power*, Cambridge: Polity.

Connell, R. (1995) *Masculinities*, Cambridge: Polity.

Connell, R. (1998) 'Masculinities and globalisation', *Men and Masculinities* 1: 3–23.

Connell, R. (2000) *The Men and the Boys*, Cambridge: Polity Press.

Dempsey, K. (1990) *Smalltown: a Study of Social Inequality, Cohesion and Belonging*, Melbourne: Oxford University Press.

Dempsey, K. (1992) *A Man's Town: Inequality Between Women and Men in Rural Australia*, Melbourne: Oxford University Press.

Fairweather, J. and Campbell, H. (1990) *Public Drinking and Social Organisation in Methven and Mt. Somers*, Research Report 207, Christchurch: Agribusiness and Rural Research Unit, Lincoln University.

Gasson, R. (1992) 'Farmers' wives and their contribution to the farm business', *Journal of Agricultural Economics* 43: 74–87.

Guardian (2003) Weekend, 12 July 2003, p. 39.

Guggenbuhl, A. (1997) *Men, Power and Myths: The Quest for Male Identity* (trans. G. Hartman), New York: Continuum.

Hearn, J. (1998) 'Theorizing men and men's theorizing: varieties of discursive practices in men's theorizing of men', *Theory and Society* 27: 781–816.

Hughes, A. (1997a) 'Women and rurality: gendered experiences of "community" in village life', in Milbourne, P. (ed.) *Revealing Rural Others*, London: Pinter.

Hughes, A. (1997b) 'Rurality and cultures of womanhood: domestic identities and the moral order in village life', in Cloke, P. and Little, J. (eds) *Contested Countryside Cultures*, London: Routledge.

Jackson, P. and Taylor, J. (1996) 'Geography and the cultural politics of advertising Progress', *Human Geography* 20: 356–71.

Jones, O. (1999) 'Tomboy tales: the rural, gender and the nature of childhood', *Gender, Place and Culture*, 6: 117–36.

Kimmel, M. (1996) *Manhood in America*, New York: A Cultural History Free Press.

Kimmel, M. and Ferber, A. (2000) 'White men are this nation: right-wing militias and the restoration of rural American masculinity', *Rural Sociology* 65: 582–604.

Law, R. (1997) 'Masculinity, place and beer advertising in New Zealand: The Southern Man campaign', *New Zealand Geographer* 53: 22–8.

Liepins, R. (1998) 'Women of broad vision: nature and gender in the environmental activism of Australia's "Women in Agriculture" movement', *Environment and Planning A* 30: 1179–96.

Liepins, R. (2000) 'Making men: the construction and representation of agriculture-based masculinities in Australia and New Zealand', *Rural Sociology* 65: 605–20.

Little, J. (1986) 'Feminist perspectives in rural geography: an introduction', *Journal of Rural Studies* 2: 1–8.

Little, J. (2002) *Gender and Rural Geography*, Harlow: Prentice Hall.

Little, J. and Austin, P. (1996) 'Women and the rural idyl', *Journal of Rural Studies* 9: 101–11.

Little, J. and Panelli, R. (2003) 'Gender research in rural geography', *Gender, Place and Culture* 10: 281–9.

Mac an Ghaill, M. (ed.) (1996) *Understanding Masculinities*, Buckingham: Open University Press.

Mackenzie, F. (1992) 'The worse it got the more we laughed: a discourse of resistance among farmers of Eastern Ontario', *Environment and Planning D: Society and Space* 10: 691–713.

Mingay, G. (ed.) (1989) *The Rural Idyll*, London: Routledge.

Morin, K., Longhurst, R. and Johnston, L. (2001) '(Troubling) spaces of mountains and men: New Zealand's Mount Cook and Hermitage Lodge', *Social and Cultural Geography* 2: 117–39.

Ni Laoire, C. (2001) 'A matter of life and death? Men, masculinities and staying "behind" in rural Ireland', *Sociologia Ruralis*, 41: 220–36.

Phillips, M. (1998a) 'Investigation of the British rural middle classes: Part 1, from legislation to interpretation', *Journal of Rural Studies* 14: 411–25.

Phillips, M. (1998b) 'Investigations of the British rural middle classes: Part 2, fragmentation, identity, morality and contestation', *Journal of Rural Studies* 14: 427–43.

Phillips, M. (1999) 'Gender relations and identities in the colonisation of rural "Middle England"', in Boyle, P. and Halfacree, K. (eds) *Gender and Migration in Britain*, London: Routledge.

Phillips, M. (2002) 'Distant bodies? Rural studies, political-economy and poststructuralism', *Sociologia Ruralis* 42: 81–105.

Phillips, R. (1995) 'Spaces of adventure and the cultural politics of masculinity: R.M. Ballentyne and the Young Fur Traders', *Environment and Planning D: Society and Space* 13: 591–608.

Rose, G. (1993) *Feminism and Geography: The Limits of Geographical Knowledge*, Cambridge: Polity.

Rose, G. (1999) 'Performing space', in Massey, D., Allen, J. and Sarre, P. (eds) *Human Geography Today*, Cambridge: Polity Press.

Rutherford, J. (1997) *Forever England: Reflections on Masculinity and Empire*, London: Lawrence & Wishart.

Sachs, C. (1983) *Invisible Farmers: Women's Work in Agricultural Production*, Totowa NJ: Rhinehart Allenheld.

Saugeres, L. (2002) 'Of tractors and men: masculinity, technology and power in a French farming community', *Sociologia Ruralis* 42: 143–59.

Smith, W. (1994) *If you haven't got any socks, you can't pull them up*. Research report to Landcare, University of Auckland (Tamaki): Division of Science and Technology.

Speight's (no year) 'Southern Man Identification Chart'. Online. Available at http://www.speights.co.nz/south_southernman.cfm (accessed 26 February 2004).

Sweetman, C. (ed.) (1997) *Men and Masculinity*, Oxford: Oxfam.

Whatmore, S. (1990) *Farming Women: Gender, Work and Family*, London: Enterprise Macmillan.

Whitehead, S. (2002) *Men and Masculinities*, Cambridge: Polity Press.

Whitehead, S. and Barrett, F. (eds) (2001) *The Masculinities Reader*, Cambridge: Polity Press.

Woodward, R. (1998) 'It's a man's life!: soldiers, masculinity and the countryside', *Gender, Place and Culture* 5: 277–300.

Woodward, R. (1999) 'Gunning for rural England: the politics of the promotion of military land use in the Northumberland National park', *Journal of Rural Studies* 15: 17–33.

Woodward, R. (2000) 'Warrior heroes and little green men: soldiers, military training, and construction of rural masculinities', *Rural Sociology* 65: 640–57.

Further reading

Campbell, H. and Bell, M. (2000) 'The question of rural masculinities', *Rural Sociology* 65: 532–46. In this introduction to the special issue of *Rural Sociology*, the authors give an excellent summary discussion on the topic 'rural masculinities'.

5

THE MALADIES OF MANHOOD IN THE BUZERPLATZ

Czech military officers in 'transition'

Hana Červinková

Summary

This chapter examines the effects of postsocialist transformation on masculinities in the Czech military. It offers an ethnographic insight into some of the dilemmas that the men who are military officers in the Czech Armed Forces face in their work for a military institution which has undergone dramatic changes in the last decade. I will show how changes related to the transition to democracy on the political level have influenced the personal and professional identity of the military men who currently find themselves at the crossroads of the old and the new military system.

- Political change and the Czech military
- Nostalgia and melancholy
- Unequal partners.

Introduction

This chapter is based on 16 months of qualitative research, which I conducted among the men and women of the Czech military in 2001–02. My goal was to learn through interviews, participant observation and the study of archival and media material, how members of the Czech Armed Forces experience the ideological and professional change from the socialist to a democratic military system. The following pages contain an ethnographic description of some of the dilemmas that the men who are military officers in the Czech Armed Forces face in their work. In my ethnographic account I wish to share with my readers the sense of how changes related to the transition to democracy on the political level have influenced the personal and professional identity of the military men who currently find themselves at the crossroads of the old and the new military system.

Political change and the Czech military

Political changes associated with the Czech transition from the socialist to a democratic system after 1989 have strongly affected the country's military, an institution which embodies the state's prerogative power and its authority over the use of legitimate violence. The extent of the effects of political changes on the country's military came into sharp focus in 1999 when the Czech Republic (together with Poland and Hungary) was admitted into the North Atlantic Treaty Organization (NATO). At this time, the Czech military came under the scrutiny of media and public attention as the country's successful integration into the brotherhood of Western democratic states began to depend increasingly on the ability of the Czech military institution and its members to adapt to a new military system. It was through the successes or failures of the Czech soldiers who participated in allied exercises and operations, that the Czech state gained or lost important points on the scale measuring the extent of the country's newly established loyalty toward NATO and the democratic world.

Given the importance which was now given to the post-socialist militaries, one would expect to encounter an abundance of activity in the military institution and a growth of pride in the military profession as such. Contrary to these expectations, however, it seemed that the institution expected to be the herald of the Czechs' victorious march into the Western world was instead an institution that seemed quite resistant to modernization and change. When I began my 16-month employment as a field researcher at a Prague-based military research institute in the winter of 2001, almost two years after the Czech Republic had gained NATO membership, the military had fulfilled only a minimum of requirements it promised to meet with its entry to the Alliance. This lack of change was a source of concern for international observers as well as the Czech political leadership, eager to use the institution for their political agenda (Gabal 2001; Gazdík 2001; Kopecký 1997; Simon 1999; Žantovský 1999). The impression that the military was actively resisting change was so strong that it led the US army attaché to say that the Czech military, whose bases are 'by and large very dumpy', instead of working to change its negative reputation, 'tends to dwell on this negative image' and attribute it to 'outside forces and not to take responsibility to improve some things on their own'. Unnecessarily, he said with disbelief and indignation, 'little things like painting the front gate and flying the Czech flag go undone' (Ulrich 1999: 147).

Such lack of initiative in the place where one expected to encounter its excess did not especially fit the historical moment. For many years, the Czech or Czechoslovak armed forces had submissively observed the course of foreign occupations of the national territory from their barracks: first the German occupation during the Second World War and then the Soviet 'temporary presence' in Czechoslovakia that lasted from 1968 to 1991. In fact, the Czechoslovak and Czech military have traditionally distinguished themselves from many other militaries by the peculiar privilege of never having fought in a war. Now, the membership in NATO seems to offer the military a chance to rescue itself from this image of impotence. But, instead of being excited about the new importance ascribed to their institution and profession, the military men I met on my travels through Czech bases did not seem particularly happy at all!

Nostalgia and melancholy

Overall, the bases on which I had lived and worked during the 16 months of my research were disconsolate places filled with an overwhelming atmosphere of melancholia. Every day, I was welcomed at the gate by young, heavily bored smoking conscripts – a dying species in the era of 'military professionalization'. These unwilling laymen, who were reluctantly fulfilling their one-year military service in the age that was to belong to enthusiastic professional practitioners of the art of military violence, made them appear irrevocably stuck in the past. The physical dilapidation of the buildings surrounding the central 'Buzerplatz', a large open space used for drilling soldiers during roll calls, combined with the obsoleteness of the omnipresent Soviet technology corresponded with the grim disposition of male military personnel. The melancholy of the military establishment seemed to grow out of the space called 'transition' that opened between the past and the future, generating a peculiar blend of nostalgias for the bygone and unattainable desires for the forthcoming.

The older male officers who had spent most of their life serving under socialism were scared for their jobs, many of which were being displaced as the military reorganized and reduced the number of personnel.[1] These men spent much time conspiring, using old-established friendships from socialist times to keep their positions as long as possible. While under socialism, people were not encouraged to leave the service and formed a life-long commitment to the military institution, the new Law on Professional Soldiers[2] passed in 1999 and effective in November 2001 (the first law on this issue since 1956), determined clear financial retirement benefits for each rank and job position, increasing with the number of years spent in service. The new law thus marked a move away from a life-long family-like commitment to the military institution,[3] toward a contractual form of obligation to the military, setting a new standard of 'the military professional'. Paradoxically, while the Czech state under the new law allotted nice pensions to the military personnel for their past service, it also provided incentives for people to stay in service as long as possible with the prospect of receiving even more money in the future. This banal reason for staying in the military, of course, was the 'public secret' within the military – a secret whose details everyone knew, but that generally remained unspoken.[4]

Instead, this older generation of men complained a great deal about what they perceived was the declining discipline within the military institution compared to the socialist times when law and order prevailed. A senior pilot I interviewed, for example, accompanied his nostalgic remembrance of the well-organized old times with an intriguing description of flight exercises:

We used to exercise all the time. There used to be the West and the East, but now you cannot call it that, because there is no enemy in the West or really in the East either. And so East and West are no longer used. Just look at the maps we now have and tell me it is not confusing. But that is bad, because even in soccer you run from one side to the other. There used to be concrete directions, there were strategic fronts. Our maps used to be marked blue, red or orange. We were red and they were blue. I do not understand what is so bad about that. Now there is no direction to fly against. We used to be on high alert all the time, there were planes ready to be flown and there was discipline and order.

In their nostalgia for past order, the former East–West divide (sometimes color-coded for easy visual recognition) manifest in the opposition between the Warsaw Pact and NATO, the older generation of officers seemed to long for a particular type of clearly conveyed and understood discipline. This discipline, which originated in the Moscow headquarters of the Warsaw Pact, was reinforced through the military structures of the alliance and conveyed in the Russian language to the militaries of Soviet colonies in East and Central Europe. During socialism, the Czechoslovak military officers thus bore a double allegiance to the nation state of Czechoslovakia, whose people they swore to protect, and to the Warsaw Pact military command through which they were subordinated to Moscow and the transnational socialist brotherhood in arms (Rice 1984; Simon 1985). During the Warsaw Pact invasion of Czechoslovakia in 1968, exposing the uncertain connection between the state and the military, the strength of transnational military command superseded the military's allegiance to the state and to the Czechoslovak civilians. By following the orders to stay in barracks and to welcome the invaders as allies, the Czechoslovak military officers also revealed the tenuous system of military allegiances and of brotherhood enforced through discipline and permeated with the threat of violence and betrayal. The Soviet, Polish, Hungarian and Bulgarian brothers in arms had betrayed the Czechs and Slovaks in a magnificent act of emasculation enacted by the latter's coerced passivity. In a quintessential homoerotic act, the invading Soviet brothers became men, while the Czechs and Slovaks who, as one officer eloquently recalled during an interview, 'opened our gates to the Soviet troops who took over our barracks', lost the fantasy of their manhood.

Was it perhaps the knowledge of this subjugation to the discipline of the friend-turned-invader, which tied the older Czech officers together in a brotherhood of military serfdom? Was it this history of Soviet *buzerace* – a vulgar word for 'ordering around', but literally meaning sodomy (*buzerant* being a derogatory term for a homosexual), which was the subject of the officers' nostalgic longing and which made institutional change difficult? The strength of fraternal military bonding among men who shared a history of conquest, would surely make the dismissal of one officer by another at a time of post-socialist military downsizing a truly fratricidal act. The predicament of this task seemed particularly clear to me when I observed the process of so-called 'civilization' of the Czech military in 2001–02. The fulfillment of the NATO requirement to increase civilian democratic control of the military by lowering the proportion of military to civilian employees was accomplished by a casual act of cross-dressing. In this ritual practiced in the Czech military since the late 1990s, the military personnel retired from the military and took off their uniform on one day, but the next returned to their jobs as civilians. With a nice retirement check, which in the case of someone who had spent 20 years in service equaled the regular monthly salary of a state employee, they returned to their job to get another monthly paycheck for the same work they now performed as civilians. Not surprisingly, according to those former military officers with whom I spoke, their major regret now that they were civilians in the military enjoying double income was a sense of mourning for the uniform, which used to bond them to their brothers in arms.

Unequal partners

The young generation of military officers I interviewed during my fieldwork grumbled as well. For their part, however, it was because they felt the military was not reorganizing efficiently and old cadres at the top echelons of the military and politicians ignorant of the needs of the military sector were keeping down their career advancement as well as the Czech military's progress. In his response to my request to speak about relations between the younger and the older generations of officers, a young pilot told me:

> The gap between the base and the headquarters has only grown. Nobody who is any good at what he does would go up to the headquarters full of old cadres who protect each other and who have no interest in changing things. Anybody who is any good at his job wants to stay down here at the base, because here you can do your work. Up there, they just create paperwork. And then, when something comes from NATO to the headquarters, they send it to us to fill out and prepare here anyway, because they have no idea what to do with it up there.

In the commentary of the young officer, the generational gap in the Czech military – according to him the reason behind the slow progress of military transition – consisted in the inability of older officers to discipline themselves according to the rules of the new transnational regime of obedience. The marker of this shift was NATO-originating paperwork, which the older officers were unable to complete in English – the new language of externally instituted military supremacy.

Obviously, however, learning English was not a sufficient 'language' skill for the passage to 'the other side'. In an interview with a star member of the elite unit in the Air Force, I was offered a startlingly clear revelation of the nature of the new regime of transnational military power:

You ask me if something changed with our entry into NATO? . . . I thought it would, but it was just a pipe dream. The guys in NATO basically view us as a different species that are still climbing trees while they themselves fly jets in the sky. That is how it is when they come here. They don't consider us their partners. And they are right. We cannot be their partners, because we really do not know anything and have nothing to offer to them.

Instead of a partnership of brothers in arms within NATO, the picture that was revealed to me in the interviews with this upcoming generation of Czech elite officers was one of the next colonial regime of *buzerace* – a new form of subordination to a transnational military brotherhood. The type of manhood, which this new regime of obedience instituted and on which it relied, however, was of a different kind than that of the past. The subordination, which was somehow perversely satisfying for Czechoslovak officers under the Soviet rule, was now a cause for a trauma and crisis of masculinity.

The self-deprecating discourse of backwardness, of lagging behind, of needing to catch up, to reorganize, to change and to transform, to finish speedy military 'transition', to be 'when and where NATO needs us', (Ministry of Defense 2002a) and yet the apparent inability to do so, has been the defining feature of discussions within the military leadership. The young Czech deputy director of strategic planning went so far as to reach into pre-human history to label the problem of change, saying that while 'the totally old dinosaurs are gone from the ranks, some problems persist' (Richburg 2002: A22). And in a propaganda leaflet, the Ministry of Defense used language from the area of nutritional and chemical instruction to identify the features of the required process of transition:

We are in NATO and we are behind in many things. . . . We are trying to be modern, but our thinking is that of old times. And we do not want to be a military unable to engage in battle, a military that only complains about the situation. Therefore – a diet. Therefore – a rejuvenating bath. Therefore – a reform. . . . First, a decorroding treatment. This means shedding the remains of old bad habits. Eliminating unnecessary, obsolete equipment and gear. And starting to think.

(Ministry of Defense 2002b)

And yet, despite great efforts, this and other campaigns for 'transition' have not found resonance in changing the melancholic attitude of most male military officers.

Contributing to the officers' sense of being behind and not at the vanguard of 'transition' is the newly emerging contrast between employment opportunities for men in the private sphere and in the state sector. This rift, previously nonexistent under socialism, seems to have a particularly strong effect on the humor of Czech military men. Once, in a moment of frustration with his private life, a very successful officer located high in the military hierarchy told me during my visit with him:

How do you think that I feel when every businessman with whom I negotiate contracts for the military arrives here in a new shining BMW, dressed
in an Italian suit, money pouring out of his ears? I command thousands of
people, have responsibility for millions of dollars, speak several languages,
but still make a miniscule amount of what this guy brings home every
month. What do you think my wife thinks about that? How can she have
any respect for me?

Such a situation of emasculation through a comparison with the emergent forms
of wealth and male power in the private sector, obviously presents a new quandary
for Czech officers. The changing gender composition of the post-socialist labor
market, where opportunities grow for men in the better-paid private sector, while
the less paying sphere of state employment is becoming dominated by women, has
important implications for the formation of gender identities.[5] The effects are even
greater in the case of military officers, who used to be rewarded for their acquies-
cence to the socialist regime with material advantages compared to people in other
state sectors, but whose former lead has since been lost. While under the last 20
years of socialism shortages in supplies plagued most institutions, the channels of
supply to the military, which remained under direct command and supervision of
Moscow, remained fully flowing. These channels of supply were completely interrupted in the 1990s, and it soon became quite clear that NATO would not readily
become the new source. And so when most military men in their interviews with
me called for the 'transition' to end, saying 'Let us already be there' (*At' už jsme
tam*), what they were also calling for were the former and anticipated times of
plenty. What they said they wanted was *koncepce* (a long-term strategic program) in
future military development, which stood for the security of past times when materials were abundant and their daily work in the military establishment proceeded
smoothly. As a senior air-ground crew leaving the Air Force told me:

For ten years they have been telling us that they are preparing a new long-
term strategic program [*koncepce*], but I cannot see one! They say things will
change but we are not getting anywhere. I have been struggling here for
ten years everyday with a lack of everything, because nothing is planned
ahead. I have no spare parts to work with – I never know what will happen
tomorrow. One cannot be sure of anything.

The destination of the officers' longing was the future imaginary place of order,
clearly defined enemy, strategy, exercise schedule, high technology, clear values and
stability – a new regime of discipline. For now, suspended in the stage of 'transition', all the unhappy military men with whom I met during my research shared a
sense of melancholic disposition, surely not a desirable trait in men whose job as
practitioners of violence in the new transnational regime of military power requires
resolve and self-assurance.

I should add that the outlook of their female counterparts confirms the importance of gender as a crucial determinant of the attitude of Czech male military officers after socialism. Revealingly, Czech military women see that which makes their
male colleagues depressed as an opportunity. While the conscription of men remains
a problem for the military, the enrollment of women is quickly rising (Vlachová and
Pávková 1997; Ministry of Defense 2002a). The 15 women officers and NCOs that I
interviewed during my fieldwork generally lacked the melancholic disposition so
characteristic of their male counterparts. When I asked them whether they were sat-
isfied with their jobs, all except one responded in one way or another that they were
quite happy with their job in the military, which provided great security compared
to the perils of the private sector. And so while the military institution in 'transition'
is an insecure and melancholy-inducing place for Czech military men, women see
the same transitional military as an unusually safe and satisfying occupational setting.

Conclusion

The melancholic disposition of the male officers in the post-socialist Czech military
seems to challenge the occupational gender model of hegemonic masculinity normally associated with men in modern military institutions. Such hegemonic masculinity, says Frank Barrett in his illuminating study of masculinity in the United
States Navy, 'refers to a particular idealized image of masculinity in relation to
which images of femininity and other masculinities are marginalized and subordinated' (Barrett 1996: 130). The military embodies and perpetuates this hegemonic
form of masculinity, characterized by such traits as perseverance, toughness, risk-
taking, technical rationality and courage and also 'plays a primary role in shaping
images of masculinity in larger society' (Barrett 1996: 129). 'The murderous hero',
violent and heterosexual, is the admired form of masculinity in Western culture
which is used by governments and other agents to mobilize support for war and hire
men into the military service (Connell 1989). Militaries around the world depend
on the definition of soldiers as embodying traditional male gender roles and thus
play a crucial role in promoting such models of manhood as a desired and hegemonic
form.

The Czech officers' melancholy, which I described in this chapter, stands in
obvious contrast to this generally accepted model of military manhood. Instead of a
hegemonic powerful masculinity associated with their membership in a military
institution, the manhood of the Czech military men is much more strongly defined
by their subservient relationship with the more powerful military brothers – previously the Soviets and now the NATO militaries. As such, the manhood of the Czech
officers whom I met resembles more closely the demasculinized form characteristic
of men in colonized cultures (Gutmann 1997; Stoler 1991). Fuelled by the uncertainty of the period of transition, saturated with nostalgic longing for the bygone and
dreaming of unattainable futures, the melancholic masculinity of the Czech military
officers challenges the uniform model of a hegemonic military manhood and points

to the complex relationship between gender identities and international political and military order.

Notes

1 The number of personnel in the Czech Armed Forces decreased from 131,965 in 1993 to 69,296 in 2001. The Ministry of Defense of the Czech Republic, *Reform of the Armed Forces of the Czech Republic*, p. 43.
2 Zákon č. 221/1999 Sb., o vojácích z povolání. Most Czech laws regarding the military can be found on www.army.cz.
3 While not specifically addressing the military, Katherine Verdery's term 'socialist paternalism' fittingly describes the quasi-family dependency of all citizens on the patriarchal socialist state, supposedly bonded by their equal share in the redistributed social product (Verdery 1994).
4 I owe the notion of the 'public secret' to Michael Taussig, who employs it in his ethnography, *The Magic of the State* (Taussig 1997: 58) and elaborates on the notion in his later work, *Defacement. Public Secrecy and the Labor of the Negative* (Taussig 1999: Part 2).
5 In their book, *The Politics of Gender After Socialism*, Susan Gal and Gail Kligman point out the growing number of opportunities for men in the private sector and the changing gender composition of different industries after socialism: 'In the newly emerging pattern of gender segregation, the heretofore feminized occupations and professions remain largely in the public sector, where labor discipline is lax, allowing an easier compromise with household obligations. In contrast, men's relative greater time flexibility, due to fewer domestic obligations, becomes a valued resource in the labor market. Young men are moving rapidly into the newly expanding, more demanding, higher-paying private sector' (Gal and Kligman 2000: 59).

References

Barrett, F. (1996) 'The organizational construction of hegemonic masculinity: the case of the US navy', *Gender, Work and Organization* 3(3): 129–42.

Connell, B. (1989) 'Masculinity, violence and war', in Kimmel, M. and Messner, M.A. (eds) *Men's Lives*, New York: Macmillan.

Gabal, I. (2001) 'Obrana – krize se prohlubuje', *Neviditelný pes*, 2 April.

Gal, S. and Kligman, G. (2000) *The Politics of Gender after Socialism*, Princeton: Princeton University Press.

Gazdík, J. (2001) 'Armádu deptají zastaralé stereotypy', *MF Dnes*, 27 March.

Gutmann, M. (1997) 'Trafficking in men: the anthropology of masculinity', *Annual Review of Anthropology* 26: 385–409.

Kopecký, P. (1997) 'Skoncujme še švejkovstvím', *Lidové noviny*, 7 March.

Ministry of Defense of the Czech Republic (2001) *Reform of the Armed Forces of the Czech Republic*, AVIS, Prague.

Ministry of Defense of the Czech Republic (2002a) *Zítřek patří profesionálům: o reformě ozbrojených sil České republiky*, AVIS, Prague.

Ministry of Defense of the Czech Republic (2002b) *Nehrajem si na vojáky: stručný návod, jak zvládnout reformu armády*, AVIS, Prague.

Rice, C. (1984) *The Soviet Union and the Czechoslovak Army, 1948–1983: Uncertain Allegiance*, Princeton: Princeton University Press.

Richburg, K.B. (2002) 'Czechs become model for new NATO: "niche" expertise in chemical weapon defense compensates for military schortcomings', *The Washington Post*, 3 November.

Simon, J. (1985) *Warsaw Pact Forces: Problems of Command and Control*, Boulder: Westview Press.

Simon, J. (1999) 'The new NATO members: will they contribute?', *Strategic Forum*, National Defense University, Institute for National Strategic Studies, Washington, DC.

Stoler, A. (1991) 'Carnal knowledge and imperial power: gender, race, and morality in colonial Asia', in di Leonardo, M. (ed.) *Gender at the Crossroads of Knowledge: Feminist Anthropology in the Postmodern Era*, Berkeley: University of California Press.

Taussig, M. (1997) *The Magic of the State*, New York: Routledge.

Taussig, M. (1999) *Defacement: Public Secrecy and the Labor of the Negative*, Stanford: University of California Press.

Ulrich, M.P. (1999) *Democratizing Communist Militaries: The Cases of the Czech and Russian Armed Forces*, Ann Arbor: The University of Michigan Press.

Verdery, K. (1994) 'From parent-state to family patriarchs: gender and nation in contemporary Eastern Europe', *East European Politics and Society* 8(2): 225–55.

Vlachová, M. and Pávková, E. (1997) *Gender v. České armádě*, Departmental Report, Prague: Ministry of Defense of the Czech Republic.

Žantovský, M. (1999) 'Můžeme být za mírotvorce, ale i za švejky', *Lidové noviny*, 11 March.

Further reading

Barrett, F. (1996) 'The organizational construction of hegemonic masculinity: the case of the US navy, *Gender, Work and Organization* 3(3): 129–42. This is a rare article in military sociology/anthropology based on qualitative research in the US Navy, which focuses on the construction of different forms of masculinity in a military setting.

Gal, S. and Kligman, G. (2000) *The Politics of Gender After Socialism*, Princeton: Princeton University Press. This book by Susan Gal and Gail Kligman is important in providing a rigorous theoretical analysis of the changing gender relations in the societies of Eastern and Central Europe in the post-socialist period.

Genet, J. (1974) *Querelle*, New York: Grove Press. The novel by Jean Genet about love and betrayal in the navy provides deep insight into the homoerotic passion that often underlies friendships among military men.

Wolff, L. (1994) *Inventing Eastern Europe. The Map of Civilization in the Mind of the Enlightenment*, Stanford, California: Stanford University Press. Larry Wolff's historical study focuses on the important, yet subordinate place of Eastern Europe in Western European imagination. Always seen as the quintessential other, Eastern Europe has played a crucial role in the construction of Europe and this place is perpetuated in contemporary practices.

Part 2

MASCULINITIES AND CULTURAL CHANGE

6

TRANSIENT MASCULINITIES

Indian IT-professionals in Germany

Bettina van Hoven and Louise Meijering

Summary

In this chapter, we discuss migration experiences of Indian IT-professionals in Germany. We describe the context of migration, in particular the motivation of IT-professionals to migrate and the anticipation of what they may encounter in the foreign country focusing on: the migrant's (economic) motives and his perception (geographical imagination) of the host country. We are interested in the conflicts that migrants may experience as a result of their desire or their obligation to move, their specific cultural baggage/difference (particularly gender, race/ethnicity, class and caste) and the way in which they encounter the different sides of German society.

- Data collection
- Culturally embedded masculinities
- Cross-cultural adaptation
- Renegotiating masculinities.

Introduction

Much research on masculinities to date has focused on white men in their own cultural context (Archer 2001). Masculinities of transients have, however, remained understudied. Transnational movements and living in a different culture does influence people's identities in general but claims about the impact on masculinities may be difficult to establish. As elements of the guest culture are adopted or parts of the home culture neglected, either by choice or by necessity, a person's understanding and performance of gendered identities is challenged. In this chapter, we use the experiences of a group of transient Indian IT-professionals in Germany to illustrate some of the conflicts they experience in performing their masculinities. The chapter is part of a larger study on the migration experiences of transient Indian IT-professionals in Germany more generally (see also Meijering and van Hoven 2003).

The findings in this chapter must be seen as tentative insights and starting points for further research on performing transient masculinities. After a short discussion of the process of data-collection, we will first describe some general cultural differences between India and Germany. This is followed by an account of how the respondents perform masculinities in India and difficulties they experience in continuing their way of life in their host country.

Data collection

As was mentioned above, this chapter is part of a larger study on migration experiences by Indian IT-professionals in Germany. In the context of the study, 22 in-depth interviews were held, all but one of which were taped and transcribed. For the analysis of the transcripts, the computer package QSR N4 was used. Although one interview was held per respondent, all respondents agreed to be approached again in case of uncertainties. All respondents were working for transnational companies (TNCs) in the Frankfurt am Main metropolitan area. A snowball approach was used to recruit the majority of the respondents; only three respondents were found with the help of official institutions, in this case the *Industrie und Handelskammer* (Chamber of Commerce) in Darmstadt. This lack of information through official channels can be related to the German privacy legislation. Consequently, it was also not possible to find respondents who were working as the only foreigner in small companies.

The above approach resulted in a sample of predominantly single males, Hindu, (upper) middle-class respondents, between 23 and 30 years old and educated in computer science or engineering. They were working for TNCs, and had been sent to the German subsidiaries by their Indian employers. Most respondents planned to spend about one and a half years in Germany. At the time of the interview, they had been in Germany for periods varying between six months and two years.

Culturally embedded masculinities

In the context of our argument, we see the performance of masculinities as embedded in culture. Culture is explained by Lewis (2002) as an 'assemblage of imaginings and meanings' (p. 13). These imaginings and meanings are expressed through language (texts) and symbols. Furthermore, they are learnt within the social environment of an individual and shared between members of a society. Therefore, culture is expressed at different levels ranging from culture bound to the family context, place specific cultures (such as the workplace), activity specific cultures (such as football), to wider cultures based on religion, ethnicity or the nation. However, it is important to note that such meanings are constructed at specific times and places and do not remain fixed. Equally, it is possible (and likely) that individuals are part of various cultures at any one time.

The Dutch sociologist Hofstede (1991) attempted to compare cultures and explain problems between them by identifying five fundamental dimensions within all cultures. In so doing, he classified and compared countries using scores on a scale from 1 to 100. Table 6.1 represents Hofstede's model as applied to India and Germany.

One of the dimensions distinguished is 'collectivism-individualism'. Indian society appears to display more collectivist features, meaning, for instance, that individuals are usually part of closely-knit groups that provide them with life-long security, in exchange for loyalty. Another dimension, 'femininity-masculinity', characterizes Indian society as slightly more feminine than German society. In India, modesty and resistance to violence are valued highly, attributes that are commonly associated with femininity. The following quote illustrates the differences between Germany and India in this respect.

> [In India, people] see to it that they don't hurt the other person . . . if, for example, you had done something wrong to me, I wouldn't put it in a way that it hurts you, I would try to convince you that what you did was wrong in a very polite manner . . . Whereas [a German] would put it in a way that [you would] understand it.
>
> (15)[1]

However, although some behaviour, or general 'traits' might be shared, not all aspects of culture are universal within a society. There are differences based on gender, social class (or caste), age groups, ethnicity, religious affiliation, to name but a few examples. In India, women taking up paid work are not usually accepted socially, especially according to ideal cultural norms in the higher castes. The ideal norm is for them to be literally restricted to the private sphere of the household, whereas men are expected to secure the incomes of their families and to move in public spaces (Chen 1995). The background of the respondents as IT-professionals – often educated in engineering – is particularly interesting in our description of

Table 6.1 German and Indian scores on five dimensions of culture

Dimension	Germany	India
Power distance	35	77
Collectivism-Individualism[a]	67	48
Femininity-Masculinity[b]	66	56
Uncertainty avoidance	65	40
Short/long-term orientation[c]	31	61

Source: Hofstede (1991).

Notes
a 100 = most individualistic.
b 100 = most masculine.
c 100 = most long-term oriented.

masculinities, since the pursuit of a scientific education and profession is seen as an important marker of masculinity in Indian society (see also Chapter 7 by Steve Derné in this volume). The status of being an IT-professional can enhance masculinities. One respondent explained:

> Your parents either wanted you to become a doctor or an engineer. Those were the things that were really high on the priority list. In India . . . you're looked at rather more with respect, if you're either from an engineering background or from [a] medical background.

(17)

The specific gender roles that appear to persist in India result in the different expectations that parents have of their sons and daughters. It is noteworthy that, in general, the behaviour and achievements of individual members of a family reflect both on themselves and on the family as a whole (Hofstede 2001). Very broadly speaking, the girls are supposed to marry a wealthy man of the right age (see also Hofstede 1998). The boys have to perform well in school and find a good job. In Indian families, men are traditionally the head of the household (extended family). Their leadership is mostly expressed in their orientation towards activities outside the home, among which is providing income. The husband is the breadwinner and he can assume that his mother, sisters and future spouse will take care of him. Men are not used to taking up domestic activities themselves. Women focus almost exclusively on care-taking activities within the confines of their household: 'I think it's unfair that the women have to . . . look after the house, . . . look after the kids . . . but in India, a lot of men get away with that' (17).

Although this division of tasks is similar if not the same as in many Western countries, in India, a woman is often literally confined to the private sphere of the home or, perhaps, one that is 'protected' by the presence of a male member of her extended family. The family honour can indeed be damaged if women appear in male spaces alone. Even in more progressive middle-class families in India, both husbands and wives acknowledge the distinction between the public and private spheres as male and female respectively. However, general norms and values, as hinted at above, are often contested by individuals. In this context, it is interesting to note that there are a few female IT-professionals who choose an international career over becoming a housewife or working in India. On the 30 November 2001, only 7.8 percent of Indian IT-professionals working in Germany were female (Leven 2001).[2] It has been found earlier that transient women more often move with their husband than alone (Hardill 1998; Yeoh and Willis 1999; Willis and Yeoh 2000). In this research, three women were interviewed. One of them was married, with her husband working in Germany as well, while the other two were single. They considered working internationally an adventurous opportunity, as well as a good career move. All three women indicated that they were very much aware of the fact that working as a transient woman was exceptional in the Indian cultural context.

The above description of restrictions to public space for women as part of the preservation of family identity is not exclusive to female family members. Men, too, comply with family customs and aim to enhance the good name of their families through their actions. For example, although a man has to attend to his own career, he is also expected to introduce other relatives in the company as part of contributing to the well-being of the family (Gupta 1999). Furthermore, children in general, but sons in particular, are expected to take care of their parents when they get older. The following quotes show the sense of responsibility that the respondents have in this regard: 'I want to stay in my country, because I have to take care of my parents. Which is a very big responsibility for me' (3). 'As [my parents] grow older, they need someone to support them. . . . They would appreciate it if I settle down in India' (15).

In this context, the concept of *Izzet* is important. It is most closely approached by the word 'status' and reflects both material well-being and experience of life (Robinson and Carey 2000). *Izzet* can be seen as a masculine quality, which can be enhanced by working abroad, because it often results in material wealth, career advancement and experiences associated with living and working in a different culture. These three issues were the most important reasons for the respondents in this study to migrate (for similar findings elsewhere see also Beaverstock 1991; Salt 1988; Stalker 2000). All three can be labelled 'masculine', since they are concerned with competition and performance (Hofstede 1998, 2001) and stress the importance of work (Levant and Kopecky 1995). Whilst 'sacrificing' a couple of years of their lives in order to enhance their careers, they increase their own status as well as the status of their families. The following quote illustrates this: 'I'm willing to sacrifice [living with my family] if I enhance my career. If I gain more money, a few years doesn't matter' (14).

The stress on financial gain by many respondents can be explained by the importance of cash as a signifier of masculinity in Indian culture. Money is considered vital in the (re-) production of Indian masculinities. Adult males, especially after returning from profitable work overseas, are expected to spend their money generously on others in order to assert their masculinity. Other men in particular will expect financial support (Osella and Osella 2000).

Cross-cultural adaptation

Based on the differences within Hofstede's (1991) cultural dimensions alone, the participation of Indian migrants in German society could be problematic. Nolan (1999), who focuses on expatriates' lives in a different culture, claims that migrants often experience a culture shock. Coping with a culture shock, he argues, can be done by 'learning your way out, and by using what you learn to change what you do' (Nolan 1999: 21). Different people tend to deal with transcultural learning and adaptation in different ways and thus adjust to their host culture to varying degrees. Riccio (2001), for example, explores a strategy by migrants whereby they neither establish boundaries nor integrate into the host culture. Instead, migrants 'live their lives simultaneously across different nation-states, being both 'here' and 'there', crossing geographical and political boundaries' (Riccio 2001: 583). Both economic resources and symbolic resources (such as goods or food from 'home') play an important part in establishing the migrants' 'transnational livelihood' (Salih 2001).

Nolan attempted to capture the various adaptation strategies in a model based on an earlier differentiation by Anderson (1994). This model (see Table 6.2) represents a typology of these different modes of cross-cultural adjustment.

When comparing the model in Figure 6.2 to the results of this research, it tran-spires that a number of variations from the model are indeed present in the research. Initially, all respondents are 'returnees' due to the limited period of their work con-tracts. To some extent, we observe withdrawal as is typical of 'escapees' and 'time servers'. Respondents with a more positive attitude toward German culture are more related to 'adjusters'. Evidently, these modes of adjustment impact upon the performance of masculinity. Prolonged contact with the host culture will create both more opportunities and restrictions for what kind of man a migrant wishes to be. Equally, negative experiences and few opportunities to be an Indian, Hindu, highly qualified man may interfere with the desire to integrate. It is notable then that Nolan's classification does not account for the gender of people in the different 'variations'. Equally, race and ethnicity seem to be of no consequence. One indica-tor of whether or not Indian males are able to integrate may be the extent to which they are able to use public and private space in ways that are familiar to them. The extent to which they recognize performances of masculinities within their host culture and the extent to which they are able to assert their own masculinities may be important.

Ways in which the Indian IT-professionals get to know their host culture are by observation and communication where possible. Usually, language forms a key factor in transmitting culture and in cross-cultural adaptation. Being unable to

Table 6.2 Model of cross-cultural adjustment

Basic mode	Variations	Description
Rejection	Returnees	The transient returns to the home-culture.
	Escapees	Transient withdraws from host-culture.
Partial adaptation	Time Servers	Minimal interaction with the host-culture, just to get through.
	Beavers	High-achievers, put much energy in keeping the host-culture at bay.
Integration	Adjusters	Positive attitudes towards the host-culture.
Biculturalism	Participators	Close involvement in the host-culture.
Multiculturalism	Learners	Learn how to learn in new cultural environments.

Source: Nolan (1999).

communicate in the language of the host-culture can prove to be an insurmountable barrier (Hofstede 1991; Benmayor and Skotnes 1994). Most respondents speak little or no German. Those who have lived in Germany longer, speak the language better, but never exceed the basic level.[3] As a result, most communication is in English, although the lack of English language skills among their German peers limits the depth of conversations. The proximity to Indian speaking peers and the restricted period of residence in Germany is decisive in the choice by most respondents not to learn the language.[4] As a result, the respondents learn about German cultural practice largely through observations of which they make meaning either by themselves or among their Indian peers.

Renegotiating masculinities

The respondents have difficulties in asserting their Indian masculinities in Germany. A key problem herein is their status as an ethnic minority with language problems. Due to their limited exposure to German culture, they get to know few German males and are therefore prevented from learning about gender roles and norms or the places where these are typically negotiated. Known ways of performing masculinity among males in public and work spaces and female family members in the private sphere are challenged in the context of their host country.

For example, because of their inability to speak and comprehend German, Indian men feel uncomfortable in public situations. As a result, they experience limited access to public spaces, which can be illustrated with the next two quotes: 'On weekends I have absolutely nothing to do [because] I don't go out anywhere' (1). 'It's basically, you really don't have much avenues where you can make friends. Like, if you go outside, . . . the language is a problem, because we don't speak German and they don't speak English' (8).

An additional factor in feeling inhibited from entering public spaces is their fear of racial discrimination. This fear is created by a combination of accounts in the media, official German policy and personal experiences. In these cases, a part of public space is claimed by a group of delinquent young males. The restrictions on access to public space make the respondents feel less secure about their masculinities.

Another space that is important for the formation of Indian masculinities is the workplace. At work, friends are made and they are also socialized with. Consequently, more time than a regular, eight-hour working-day is spent there. Furthermore, working hard contributes to the status of a man. If projects need to be completed, overtime is invested in getting the job done. Even though work is a place for social contacts, the respondents are very focused on their work.

In Germany, habits concerning work are different. According to the respondents' observations, Germans tend to work eight hours per day, even in busy periods such as when several projects need to be completed. The German co-workers made a point of returning home to their families on time, stressing that there is more to life than just work (see also Levant and Kopecky 1995). Some respondents value this approach:

> In India . . . the task is for thirty days, you have to finish that task. So here, I find it OK, because you're not that stressed really. Whereas the work goes well. And you have more free time. I like it here more.
>
> (9)

The respondents also noted that Germans do not socialize much with their colleagues. They prefer to return home to spend time with their family, or to meet close friends. However, in contrast to their experiences in India, public spaces do not play a significant role in this. German social life is perceived as very much planned and organized, where Germans make appointments rather than visiting each other spontaneously, as is common in India:

> Then, it is with Germans that a formality is involved, [whereas] with Indians, we do not have any formality. My friends just come home anytime and have dinner with me. But I have to take an appointment if I have to go to a German colleagues' house. It's mainly because of the formality.
>
> (3)

Although family life is very important for the respondents, as was illustrated above, they perceived the freedom of married men to be much higher in India than in Germany (see also Chen 1995). In India, men have few responsibilities in the household, while their wives take care of household and children. In Germany these days, men are expected to participate more actively in their nuclear families. Two respondents explain: 'I mean things like marriage, relationships, family . . . These things are best in India and what I see here, I cannot accept' (14).

> In India it's not like that, . . . sometimes it's unfair, because there are some men who have families, [but] they hang around with their friends. They don't pay too much attention to their family. . . . It's not like you get divorced if you don't. . . . But here, they have a lot of commitments here.
>
> (17)

Overall, the above factors do not encourage the IT-professionals in this study to socialize with German men whom they perceive as relatively 'closed'. They are unable to become real friends with the Germans and to integrate in a German circle of friends who meet either at home or in the pub. An additional problem in this is the fact that, due to their religion, many respondents do not drink beer or eat meat, both of which are thought by Indian males to be symbols of German masculinity. '[Germans] might feel that we have some [restrictions], they might feel that without a drink, it could not be [fun] or they might dislike Indian food.' (11)

It would seem that neither the usual places for Indians to 'be men' are experienced as easily available to them in the respondents' host country, nor are German males perceived as easy to mix with. Unless the respondents are able and willing to let go of a part of their cultural identity, they are forced to retreat to the spaces of their private homes, i.e. spaces of the feminine where they can celebrate their Indianness even without the presence of women (see Chapter 7 by Steve Derné in this volume). From the interviews, it emerged that this is indeed what many respondents do: 'Back in India, when you stay with the family, it's normally the mother who does the cooking. [Here] I spend a lot of time cooking on weekdays, because that's the best [way of passing time] for me' (1).

Once in Germany, they have to take up the feminine tasks of cooking and cleaning themselves. Furthermore, home is the place to socialize with Indian peers. In theory, these social contacts can help the respondents perform familiar ways of being men. However, due to the desire to cultivate symbolic resources from India to construct their personal homes in the foreign culture (see also Salih 2001), an important part of socializing consists of the preparation and sharing of Indian meals, a traditionally feminine task. 'We do cook [at night], so that we can have some Indian food' (8).

Conclusion

Performing masculinity depends on opportunities and constraints. Interaction with other males and females is important in experiencing what kind of man one is. In India, much interaction with other males takes place in the public and work space. At home, the man is pampered by his mother or wife. His greatest obligation is to enhance the good name of the family. As a part of this, a work assignment abroad with the prospect of career advancement, a good income and a better social status is desirable. Although in theory German males may experience their manliness in similar ways, there are differences that make it difficult for Indian males to adjust.

In German public places, the respondents to this study feel restrained by their limited knowledge of the German language and a fear of racial discrimination. Both at work and in public places they find it hard to establish contacts. They perceive German colleagues as less inclined to engage in social activities than other Indians. The respondents also find the formal nature of establishing German masculinities difficult, i.e. by appointment either at home or the pub, rather than at work or in public spaces. Consequently, they start to redefine their masculinities at home. This can be seen as an attempt to maintain their Indian masculinities in a different context and can be related to limited possibilities and motivations for cross-cultural adjustment. The more convenient new place for their masculinities happens to be a typically feminine place in India. Although some respondents felt uncomfortable with this redefinition, they all highly valued the maintenance of Indian traditions, even if at home. Symbolic resources, in particular the preparation and sharing of Indian meals, are important mechanisms for maintaining a culturally familiar and safe space.

Upon returning to India, the respondents will probably have achieved their initial goals: improving their *Izzet*, through working on their career, gaining extra money and becoming more experienced as a person. In order to achieve these goals, they have temporarily 'moved' performing their masculinities from public space to the explicitly feminine space of the home. It remains to be seen to what extent their 'new' masculinities will persist in India. The renewed accessibility and familiarity of public space and peer pressure to go out (and pay) will perhaps prevent them from remaining at home.

Notes

1 The number following the quote refers to the number of the respondent allocated in this research for reasons of confidentiality.
2 The figures only concern IT-professionals who are working in Germany on a green card, which is a specific type of visa. Holders of either a short- or long-term visa are not included.
3 Except for three respondents who had either studied in Germany or studied German in India.
4 Many acknowledged that, in order to establish contacts with 'average' Germans, good knowledge of the German language was indispensable.

References

Anderson, L.E. (1994) 'A new look at an old construct: cross-cultural adaptation', *International Journal of Intercultural Relations* 18(3): 293–328.

Archer, L. (2001) 'Muslim brothers, black lads, traditional Asians: British muslim young men's constructions of race, religion and masculinity', *Feminism and Psychology* 11(1): 79–105.

Beaverstock, J.V. (1991) 'Skilled international migration: an analysis of the geography of international secondments within large accountancy firms', *Environment and Planning A* 23: 1133–46.

Benmayor, R. and Skotnes, A. (eds) (1994) *Migration and Identity*, Oxford: Oxford University Press.

Chen, M. (1995) 'A matter of survival: women's rights to employment in India and Bangladesh', in Nussbaum, M. and Glover, J. (eds) *Women, Culture and Development*, Oxford: Clarendon Press.

Gupta, R.K. (1999) 'The truly familial work organization', in Kao, H.S.R., Sinha, D. and Wilpert, B. (eds) *Management and Cultural Values: The Indigenization of Organizations in Asia*, New Delhi: Sage Publications.

Hardill, I. (1998) 'Gender perspectives on British expatriate work', *Geoforum* 29: 257–68.

Hofstede, G. (1991) *Cultures and Organisations: Software of the Mind*, London: McGraw-Hill.

Hofstede, G. (1998) 'The cultural construction of gender', in Hofstede, G. (ed.) *Masculinity and Femininity. The Taboo Dimension of National Cultures*, Thousand Oaks, London and New Delhi: Sage Publications.

Hofstede, G. (2001) *Culture's Consequences: Comparing Values, Behaviors, Institutions, and Organizations across Nations* (2nd edn), Thousand Oaks, London and New Delhi: Sage Publications.

Levant, R.F. and Kopecky, G. (1995) *Masculinity Reconstructed. Changing the Rules of Manhood – At Work, in Relationships and in Family Life*, New York: Penguin Books.

Leven, B. (2001) 'Statistical data on green card immigrants'. Online. Available at http://Bonn-ZAV.IT-Experts@arbeitsamt.de (accessed 25 February 2004).

Lewis, J. (2002) *Cultural Studies: The Basics*, London: Sage Publications.

Meijering, L. and Hoven, B. van (2003) 'Imagining difference. The experiences of 'transnational' Indian IT-professionals in Germany', *Area* 35(2): 174–82.

Nolan, R.W. (1999) *Communicating and Adapting Across Cultures: Living and Working in the Global Village*, Westport: Bergin & Garvey.

Osella, F. and Osella, C. (2000) 'Migration, money and masculinity in Kerala', *The Journal of the Royal Anthropological Institute* 6(1): 117–34.

Riccio, B. (2001) 'From "ethnic group" to "transnational community"? Senegalese migrants' ambivalent experiences and multiple trajectories', *Journal of Ethnic and Migration Studies* 27(4): 583–99.

Robinson, V. and Carey, M. (2000) 'Peopling skilled international migration: Indian doctors in the UK', *International Migration* 38(1): 89–108.

Salih, R. (2001) 'Moroccan migrant women: transnationalism, nation-states and gender', *Journal of Ethnic and Migration Studies* 27(4): 655–71.

Salt, J. (1988) 'Highly-skilled international migrants, careers and internal labour markets', *Geoforum* 19: 387–99.

Stalker, P. (2000) *Workers Without Frontiers: The Impact of Globalisation on International Migration*, Boulder: Lynne Rienner Publishers.

Willis, K.D. and Yeoh, B.S.A. (2000) 'Gender and transnational household strategies: Singaporean migration to China', *Regional Studies* 34: 253–64.

Yeoh, B.S.A. and Willis, K.D. (1999) '"Heart" and "wing", nation and diaspora: gendered discourses in Singapore's regionalisation process', *Gender, Place and Culture* 6: 355–72.

Further reading

Bryceson, D. and Vuorela, U. (2002) *The Transnational Family: New European Frontiers and Global Networks.* Oxford: Berg Publishers. This publication offers in-depth information on how immigrants face troubling issues of cultural identity, economic change, political uncertainty and social welfare. It covers a historical perspective and a wide range of cases.

Jackson, P., Crang, P. and Dwyer, C. (eds) (2004) *Transnational Spaces*, London: Routledge. This new book focuses on space rather than transnational migrants and is a good addition to the literature used for this chapter.

Vertovec, S. and Cohen, R. (eds) (1999) *Migration, Diasporas and Transnationalism*, Cheltenham: Edward Elgar Publishing. The articles in this volume of the International Library of Studies on Migration represent key contemporary theories, comparative research and case studies on post-migration communities.

7

GLOBALIZATION AND MASCULINE SPACE IN INDIA AND FIJI

Steve Derné

Summary

This chapter discusses the effects of global media representations on Indian masculinities. Indian men often find themselves attracted to foreign lifestyles which are celebrated in global media. This attraction may destabilize men's sense of identity. Men often address their anxieties by emphasizing a male identity rooted in men's free use of cosmopolitan spaces. Nationalist Indian men who were attracted to Western forms of economy and statecraft found their Indianness in Indian women's restriction to the home. Contemporary Indian men who are attracted to cosmopolitan fashion find their identity by contrasting their free use of cosmopolitan spaces with women's home-based religious and familial duties.

- Indian men's nationalism
- Contemporary male Indian filmgoers
- Fiji Indian men respond to transnational cultural flows.

Introduction

Transnational movements of people and mass media heighten people's awareness of different lifestyles (e.g. Appadurai 1996). Hollywood productions like *Beverly Hills 90210* and *The Bold and The Beautiful* are widely watched in countries from Fiji to Turkey, Nigeria to India. Villagers in India, Fiji, and elsewhere interact with village friends who live or have lived abroad. Migration and mass mediation allow people to imagine a wider range of possible lives, some of which include greater freedom for women in cosmopolitan public spaces. Transnational media produce a desire for consumer lifestyles (e.g. Mankekar 1998; Thussu 2000). But awareness of alternative gender orders can destabilize notions of masculinity and femininity which root identity, while the celebration of seemingly foreign consumer lifestyles can threaten national identity.

Focusing on Indian men in India and in Fiji, this chapter argues that men have often handled this attraction to lifestyles, which may destabilize their masculine and national identity, by rooting their maleness in their free use of cosmopolitan spaces. They find their Indianness in women's restriction to domestic spaces of kitchens, temples, and cultural performance. These men end up rooting their own national identity in what they imagine women do in domestic spaces (into which men attempt to make only limited incursions), while defining their masculinity in terms of their free and relatively exclusive use of public spaces, like commercial enterprises and movie halls. While introducing new possibilities, then, globalization has ironically reinvigorated Indian men's commitment to spatial arrangements that are segregated by gender.

I illustrate this argument by applying it to three cases. First, I argue that nationalist Indian men in the fight for independence found their masculinity in competing in the public arena. They handled their ambivalence about embracing seemingly alien Western ways by celebrating Indian women who cooked and fulfilled family needs in the home as men believed women always had. Second, I argue that contemporary men in India, who often exhibit their masculinity by parading their cosmopolitan lifestyles in public, find their Indianness by emphasizing their attraction to Indian women who stay restricted to domestic spaces. Third, I argue that contemporary Indian men in Fiji, who are attracted to lifestyles that are celebrated in Hollywood productions, root their Indianness in women's actions at home and in the temple. I conclude by arguing that these men often especially focus on the dangers women face in the globalized public sphere, reserving increasingly privileged transnational spaces for themselves.

Indian men's nationalism[1]

From the latter half of the nineteenth century, some Indian men's focus on succeeding in the public realm by following principles they believed to be introduced by a foreign power generated anxieties about identity that led them to emphasize an Indianness rooted in women's restriction to private space. Partha Chatterjee's (1989: 623) groundbreaking work demonstrates that nationalist Indian men who struggled against colonialism were attracted to 'science, technology, rational forms of economic organization [and] modern methods of statecraft', which they believed 'had given the European countries the strength to subjugate the non-European people and to impose their dominance over the whole world'. Hence, Indian nationalists believed that to overcome European domination, they had to learn the colonialists' 'superior techniques of organizing material life'. But nationalists feared that imitating the West threatened the 'self-identity of national culture'. Chatterjee (1989) argues that nationalist Indian men handled their anxiety about embracing seemingly new political and economic structures by emphasizing Indians' superiority to Westerners in the 'spiritual domain'.

Because nationalist Indian men identified the public world as a typically male domain, they made women the primary bearers of Indianness in the crucial spiritual domain: 'The home was the principle site for expressing the spiritual quality of the national culture, and women must take the main responsibility for protecting and nurturing this quality' (Chatterjee 1989: 626–7). Hence, while men's work in the public domain required them to make:

> changes in dress, food habits, religious observances and social relations, [e]ach of these capitulations now had to be compensated by an assertion of spiritual purity on the part of women. Women must not eat, drink, or smoke in the same ways as men; they must continue the observance of religious rituals which men were finding difficult to carry out; they must maintain the cohesiveness of family life . . . to which men could not now devote much attention.
>
> (Chatterjee 1989: 629)

In short, to handle the threats to their own Indianness, nationalist men insisted that women remain at the home cooking, doing *puja* [religious observances], and maintaining family ties. Gandhi (cited by Patel 1988: 381) tried to involve women in the independence movement but he did so by focusing on their domesticity, urging that women facilitate India's self reliance by producing goods within the home. He insisted that there was 'no need for women . . . to go to work . . . for the sake of [a few] rupees'. Women, Gandhi said, 'have plenty of work in their own homes . . . They may give peace to the husband when he returns home tiredly, minister to him, soothe him if angry and do any other work they can staying at home' (Gandhi cited by Patel 1988: 381). Male nationalists apparently identified feminine domesticity and religiosity as a sign for nation.

Contemporary male Indian filmgoers[2]

In 1991 and again in 2001, I conducted research with young male Indian filmgoers in the small city of Dehra Dun, India. Each day over the course of three months in each year, I carried out observation in the eight movie houses in Dehra Dun, participating in filmgoing rituals and watching people watch films. I interviewed young men in their teens and twenties[3] with good standards-of-living but who did not speak any (or very much) English. While the interviews were not systematically random, I approached men who were purchasing expensive balcony tickets and cheap floor tickets at all of the theatres in Dehra Dun and at showings of hit films and run-of-the-mill films, fighting-and-killing films and love stories, Hindi films and English language films. The rate of refusals was low. The interviews were conducted in Hindi and took about 30 minutes. The taped interviews were conducted in public in and around cinema halls. In both years, I conducted the interviews with the help of a journalist for a regional Hindi newspaper (see Derné 2000a for further description of the interview process).

My interviewees were professionals or successfully self-employed people (23

Figure 7.1 The metal gates have just opened, if slightly, and Hindi filmgoers press into a queue to enter a movie theatre showing *Swarag Yahan, Narak Yahan* (*This is heaven, This is Hell*) (*source:* Steve Derné).

percent in 1991; 16 percent in 2001), undergraduate or postgraduate students (41 percent in 1991 and 50 percent in 2001), and successful laborers or holders of lower-middle class jobs such as office clerk (36 percent in 1991, 34 percent in 2001). Their families owned scooters or television sets, but they could barely dream of owning automobiles or traveling abroad. In 1991, male filmgoers were aware of foreign lifestyles, but by 2001 they were bombarded with new media: while none had seen cable television or Hollywood films in 1991, by 2001 two-thirds of the men I interviewed sought out foreign media. Despite the new media celebrating greater freedoms for women, the men whom I interviewed in 2001 were even more insistent on restricting women to the home (Derné 2003, 2004, forthcoming).

Sanjay Srivastava (2001: 241) shows that the postcolonial masculinity constructed in Indian films in the 1950s and 1960s emphasized 'the Five Year Plan hero' where 'manliness came to attach not to bodily representations or aggressive behaviour but, rather, to being "scientific"'. 'Roads . . . highways and metropolitan spaces came to be the "natural" habitat of the hero' who quite often was an engineer, doctor, scientist or bureaucrat. Indian men today continue to be attracted to a bureaucratic political economy and identify the public realm as their own sphere.

Many Indian men value the movie hall as a public realm in which they can show off their masculinity. Men shout and dance during popular dance and fight scenes. They seem to particularly enjoy horseplay and joking with their male friends.

At screenings of run-of-the-mill fighting-and-killing films, men usually constitute more than 90 percent of the audience. While women may make up a quarter or more of the audiences for hit social films, they tend to be segregated in the more expensive balcony seats, half of which are often reserved for women. The cheaper seats are often almost exclusively male (as is the men's half of the balcony), facilitating men's shouting and dancing, which can include harassment of women who intrude into the male spaces of the theatre (see Derné 2000b, 2003, 2004, forthcoming).

One married 25-year-old whom I interviewed in 2001 illustrates this emphasis on men's exclusive use of the cinema-hall public sphere. He describes his compelling attraction (*chasaka*) to Hindi film as so strong that he used to see at least one movie a day. Even now, he says, he sees at last two movies a week. He seems proud, however, that his wife of seven years is less interested in filmgoing. She does not like the 'seductive dresses of the heroines', he says. She is so 'home loving' (*gharelu*) that 'she even objects to seeing movies with her own husband'. For this man, the cinema hall is a place that men alone enjoy. Women should remain at home and men should avoid the domestic. This man does not watch television at home much. He says: 'It's for women and children.'

Through filmgoing and satellite television, many men have become attracted to styles that originate in Hollywood. Increasingly, filmmakers are using Hollywood fashions they see on *The Bold and the Beautiful* and *Beverly Hills 90210* to give films a glamorous, hip, glossy feel (e.g. Chopra 1997). Throughout the 1990s, Hindi films exaggerated Western name brand clothing (e.g. Levi's, DKNY) worn by heroes. Indian-film fan magazines heighten Indian-film fans' awareness of and fascination with lifestyles associated with the West (e.g. Appadurai 1996). Magazines with names like *Stardust* and *Tinseltown* include all sorts of non-Indian expressions like 'abracadabra', 'razzmatazz', and 'ooh la la'. Fan magazines include sections on American films and fan-magazine advertisements often feature Indians in Western-ized clothes.

For the great majority of filmgoing men in cinema halls, the high-flying cosmopolitan world of automobiles and jet travel portrayed in film is beyond reach, but they still appear to be attracted to Western lifestyles. Despite the lack of cosmopolitan opportunities that is typical in urban India, male filmgoers neverthe-less show a fascination with cosmopolitan lifestyles. Some sport westernized clothes, like Nike T-shirts, Rayban sunglasses, and shirts with English messages (like 'Because I'm the Best that's Why'). More than a few filmgoers have American flags sewn onto their pants or shirts – and fascination with American lifestyles appears to be increasing (Chakravarti 1998). One 23-year-old college student whom I inter-viewed in 1991 told me that he sees films partly to educate himself 'about what it means to be modern [English: modern]'. A 22-year-old high school graduate who migrated from the village to the city and whom I interviewed in 2001 told me that 'watching films' has been teaching him 'the ways of life in the city'.

Embracing fashions they see on the screen, men in and around cinema halls often

comb, style, and check their hair in the many flashy mirrors that cinema hall propri-
etors have set up in cinema hall lobbies and grounds. They enjoy 'promenading in
public with no other purpose than to show off' (Alter 1992: 242). Yet because identification with Indian tradition is an essential part of many Indians' self-conceptions,
Indian men also often feel anxiety about their attraction to Western cultures. While
men often embrace Western lifestyles, they are upset by the (often imagined) movement of women into the public realm. Men's belief that modern influences will
jeopardize women's completion of housework and childcare in the private sphere is
the centerpiece of their critique of Western influences. Phoolchand Mishra, a 35-
year-old, whom I interviewed in 1987, complained, for instance, that today women
have a craze (*phitur*) for moving around (*ghumna*), attending parties, and being
independent. Phoolchand condemned modern women who want to move in public
'standing shoulder to shoulder with men' for forgetting that children need their
mothers at home 24 hours a day. Many men similarly complained that an educational system influenced by foreign principles had made women unwilling to do the
domestic work needed in the private sphere (Derné 1994).

While films celebrate Western lifestyles, they also help Indian men handle their
ambivalence about Western ways by celebrating Indian women's devotion to home-
based family duties. Men identify their Indianness with their attraction to such
women, allowing them to more comfortably embrace elements of modernity – like
Western clothing styles and working in a bureaucratic economy – that would otherwise seem opposed to Indian traditions.

The most popular Indian films since the late 1970s contrast traditionally modest
Indian heroines with 'Westernized' Indian women who are represented as immodest
and overly forward (Derné 2000b; Thomas 1995; Uberoi 1998; Mankekar 1999). In
the 1989 influential hit movie *Maine Pyar Kiya* the heroine is a Suman, a village girl
whose devotion to family is emphasized as Indian by the contrast the film draws
between her and the excessively modern Sima. The audience first sees a cigarette-
smoking Sima drag-racing through the public streets as she plots to seduce the hero
in order to control his father's commercial enterprise. Immediately upon the hero's
return from America (a prime symbol of modernity), Sima tries to seduce him with
her short hair and revealing Westernized clothes. The hero rebuffs her advances,
rejecting a woman who is presented as excessively modernized. This contrast
between the Westernized Sima (who manipulates within the public realm) and the
traditional Suman (who emphasizes family obligations) is repeated throughout the
film. After becoming frustrated when the modest Suman gives priority to a domestic
request from the hero's mother over the hero's request that she go boating with him
in public, the hero tries to torment Suman by telling his mother that he dreams of
marrying a modern woman: 'I want jeans and a short mini skirt. I don't want long
hair, but very short hair.' This amuses his mother who laughs heartily, making it
clear that such a woman would be unsuitable for marriage because she would neglect
family duties. The hero's mother questions whether such a girl could respect elders
in the family, treat people her own age as if they were her own, and love the house's

younger people. Laughter at the thought of marrying a modern woman discomforts the hero as he storms out.

All of the men whom I interviewed in Dehra Dun in 1991 praised modest hero-ines like Suman for their commitment to family duties. An 18-year-old high school student said, for instance, that he liked 'the role taken by Suman because she serves her family in a good way. She is one who understands'. A 20-year-old said that 'a modern woman like Sima will not be able to do the good work of serving the family'. Men in India who are unable to hitch their dreams to the global economy continue to insist on women's restrictions to the home. Abraham's (2001) inter-views from 1996 to 1998 show that most low-income male college students want women to be simple, home-loving, and compromising. Similarly, Umesh, a civil draftsman whom I interviewed in 2001, complained that Satellite TV wrongly teaches a sister to 'go with her boyfriend to watch a movie'. In response to Valen-tine's Day celebrations, which have grown due to cable music channels and card marketing, protesters have attacked couples in restaurants (Sengupta 2001). Other protesters targeted discotheques for 'spoiling the minds of youth' (*India Abroad* 1999) and rallied against the 1996 staging of the Miss World pageant in Bangalore, arguing that it threatened Indian womanhood (Oza 2001; Fernandes 2000). The uneasiness of men like Umesh, which is reflected in protests against couples appear-ing in public and women appearing in beauty pageants, shows many men's attach-ment to monopolizing the public sphere as a male arena. Filmgoing men, who find their masculinity in use of the public sphere, handle their ambivalence about their attraction to Western lifestyles through their commitment to keeping women restricted to the private realm. Both their sense of manhood and their sense of Indi-anness depend on it.

Fiji Indian men respond to transnational cultural flows[4]

Indian men in Fiji, who are exposed to even more transnational cultural flows than men in India, also appear to handle conflicts between Indian and cosmopolitan iden-tities by rooting their Indianness in their own use of public space and their restric-tion of women to the domestic space of the home. Constituting about 42 percent of Fiji's population today, Fiji Indians are mostly descended from indentured laborers brought from India by the British colonial government between 1878 and 1919 to grow sugar cane.

Today, Fiji Indians encounter transnational cultural flows from both India and the US. Hindi films are routinely shown in Fiji's cinema halls; Indian religious leaders preach in Fiji; the India-based *Arya Samaj* (a Hindu organization) runs most schools that Fiji Indians attend; and Indian film stars give concerts in sold-out stadiums. But Fiji Indians also are exposed to cultural flows from the Western centers: Hollywood and New Zealand productions like *Beverly Hills 90210*, *Friends*, *ER*, *Veronica's Closet*, and *Xena* dominate Fiji television programming and Hollywood productions screen in more theatres than Bombay film productions. In 1998, flyers which announced the time of a Fiji Indian student association meeting as 'after *Xena*', showed that such television shows are a routine part of many Fiji Indians' lives. Fiji Indian men show a strong attraction to Western culture. Some want to immigrate to the US and many wear T-shirts with Western slogans.

One way in which Fiji Indian men emphasize Indianness is by focusing on what women do in the private sphere. In talking about their Indianness, the men whom I interviewed often talked about the distinctive tradition of eating curry. Events of the Fiji Indian student association usually emphasized preparation of traditional food. But men rarely cook the food. The annuals of the Fiji Indian Students' Association often depict women cooking curries. Fiji Indian men also talk about doing *puja* as an important sign of Indianness, and the annuals of the Fiji Indian Students' Association emphasize religious celebrations. Norton (2000) reports that Fiji Indians affirm pride in religion through Hindu festivals and the continued construction of Hindu temples. Again, however, worship at the temples that I visited in Fiji was primarily carried out by women and the Annuals of the Fiji Indian Students' Association show women performing religious rituals. Fiji Indian men also seem to associate their own Indianness with women's domestic location. A joke published in an Annual of the Fiji Indian Students' Association associated excessive Westernization with men's work in the domestic realm:

> A couple who emigrated to the United States several years ago [did so] with one dream – to become citizens. Though [sic] much red tape and years of study they were patient and hopeful. Then one day the husband rushed into the kitchen with the long awaited good news, 'Lizzy!' he shouted. 'At last we are Americans!' 'Fine', replied the wife, tying her apron around him. 'Now you wash the dishes'.

As in India, then, Fiji Indian men, too, may root their Indianness in women's work in the domestic sphere. Men may wear Western clothes, watch *Xena*, and dream of living in the US, but they are still essentially Indian because of their attraction to traditionally modest women who faithfully meet religious and family duties at home. A man only abandons his fundamental Indianness if his wife makes him do the dishes.

Conclusion

A similar pattern emerges in each of the three sites of Indian culture examined here. For nationalist Indian men, filmgoing Indian men, and diasporic Indians in Fiji, increasing transnational cultural flows heighten men's attraction to cosmopolitan lifestyles. While men are often attracted to cosmopolitan lifestyles, transnational media nonetheless challenge men's gender and national identity. In responding to these challenges, men increasingly emphasize the use of public cosmopolitan spaces as essential to their masculinity, while focusing on women's actions in domestic and religious spaces as the basis of their national identity.

It appears, moreover, that this reinvigoration of segregated space is a common response to the new possibilities and anxieties created by globalization. Zimmer-Tamakoshi (1993: 61–2) reports, for instance, that in Papua New Guinea the 'nationalist male elite' who participate in a translocal economy 'condemn women for striking out in modern directions . . . that many have permitted themselves'. Although they have embraced city living and participation in a wider economy, these men celebrate traditional Papuan women whom they construct as 'chaste and self-less', criticizing women who work in cities as 'Westernized', 'sexually promiscu-ous', and 'living in selfish abundance'. Ong (1999: 152) similarly focuses on how overseas Chinese men handle the 'anxiety engendered' by the large numbers of women in the paid labor force by developing 'twinned ideologies of feminine domesticity and masculine public life'. Absent the possibility of territorial-based nationality and influenced by transnational media produced by ethnic Chinese, over-seas Chinese increasingly identify 'family and female self sacrifice' as 'symbols of national unity' (Ong 1999: 154, 162).

Indeed, it appears that as prestige increasingly attaches to transnational move-ment, Indian men define transnational movement as male while seeing restrictions on movement as female. A number of recent Hindi films focus on non-resident Indians (NRIs) who have succeeded in England, Europe or New Zealand. In these films, the West is still presented as the 'site of rampant sexuality and promiscuity' (Mankekar 1999; Uberoi 1998), but NRIs are able to remain Indian by resisting these temptations. The NRI heroines of these films obey family hierarchies and avoid liquor (e.g. Mankekar 1999; Uberoi 1998), while the male heroes do not take sexual advantage of Indian or Western women and respect their obligations to their parents (e.g. Mankekar 1999; Uberoi 1998). Mankekar (1999: 737) suggests that these films portray the West as a 'place where cultural purity and authenticity *can* be maintained'. But it is significant that the rampant sexuality of Westerners puts 'the purity of NRI women . . . always at risk'. 'The West is a place where the virginity of Indian women is under constant threat' (Mankekar 1999: 739) and (in films) NRI men protect Indian women from these assaults (ibid.). Mankekar's analysis of these films emphasizes that 'the Indian nation . . . is where Indian women's sexual purity may be preserved. Indeed, the sexual purity of Indian women becomes iconomic of the purity of the national culture'. The threats and temptations that women face in Western settings are, then, far more consequential, in men's eyes, than threats and temptations that men themselves face. While these films focus on NRI women who *are* able to maintain their Indianness, their effect is to emphasize the unacceptable risks that NRI women face in transnational spaces. It appears, in short, that as trans-national movement is increasingly valued, Indian masculinity is associated with such movement, while Indian femininity is defined in opposition to it. Men may see par-ticipating in globalization as masculine, while insisting that women limit similar par-ticipation. Under the influence of globalization, Indian men may find maleness in movement throughout transnational spaces while finding their Indianness in women's restriction to enclosed domestic spaces.

Notes

1 This section draws on Derné 2000a.
2 This section is primarily based on interviews and participant observation that I conducted with filmgoing men in Dehra Dun, North India in 1991 (see Derné 2000b), but I also draw on a replication of this study which I conducted in Dehra Dun in 2001 (Derné 2003, 2004, forthcoming). The findings discussed in this section were similar in both years. I also include several examples from interviews about family life that I conducted with Indian men in 1987 (see Derné 1995).
3 In 1991, 91 percent (20/22) of the men whom I interviewed were in their teens or twenties and 86 percent (19/22) were unmarried. In 2001, 94 percent (30/32) men I interviewed were in their teens or twenties and 84 percent (27/32) were unmarried.
4 This section is based on preliminary research that I conducted in Fiji over four weeks in 1998. I conducted 13 open-ended interviews about identity with men in the sugar town of Labasa and the cosmopolitan city of Suva. In addition, I rely on publications of an association of Fiji Indian college students (see Derné 2002 for detailed references).

References

Abraham, L. (2001) 'Redrawing the *Lakshman Rekha*: gender differences and cultural constructions in youth sexuality in urban India', *South Asia* xxiv: 133–56.

Alter, J.S. (1992) *The Wrestler's Body: Identity and Ideology in North India*, Berkeley: University of California Press.

Appadurai, A. (1996) *Modernity at Large: Cultural Dimensions of Globalization*, Minneapolis: University of Minnesota Press.

Chakravarti, S. (1998) 'Americana: like, this is it!', *India Today*, 9 February: 42–5.

Chatterjee, P. (1989) 'Colonialism, nationalism, and colonized women: the contest in India', *American Ethnologist* 16: 622–33.

Chopra, A. (1997) 'Bollywood: bye-bye Bharat', *India Today*, 1 December: 53–4.

Derné, S. (1994) 'Arranging marriages: how fathers' concerns limit women's educational achievements', in Mukhopadhyay, C.C. and Seymour, S. (eds) *Women, Education, and Family Structure in India*, Boulder: Westview.

Derné, S. (1995) *Culture in Action: Family Life, Emotion, and Male Dominance in Banaras, India*, Albany: SUNY Press.

Derné, S. (2000a) 'Men's sexuality and women's subordination in Indian men's nationalisms', in Mayer, T. (ed.) *Gender Ironies of Nationalism: Sexing the Nation*, New York: Routledge.

Derné, S. (2000b) *Movies, Masculinity, and Modernity: An Ethnography of Men's Filmgoing in India*, Westport Ct: Greenwood Press.

Derné, S. (2002) 'Globalization and the reconstitution of local gender arrangements', *Men and Masculinities* 5: 144–64.

Derné, S. (2003) 'Arnold Schwarzenegger, Ally McBeal, and arranged marriages: globalization's effect on ordinary people in India', *Contexts* 2(1): 12–18.

Derné, S. (2004) *Globalizing Gender Culture: Transnational Cultural Flows and the Intensification of Male Dominance in India*, Occasional Papers of the Office of Women's Research, Manoa: University of Hawaii, 1–50.

Derné, S. (forthcoming) 'Globalization and the emergence of a transnational middle class: implications for class analysis', in Robinson, W.I. and Applebaum, R. (eds) *Critical Globalization Studies*, New York: Routledge.

Fernandes, L. (2000) 'Nationalizing "the global": media images, cultural politics and the middle class in India', *Media, Culture & Society* 22: 611–28.

India Abroad (1999) 'Ban sought on discotheques', 19 March: 11.

Mankekar, P. (1998) 'Entangled spaces of modernity: the viewing family, the consuming nation and television in India', *Visual Anthropology Review* 14: 32–45.

Mankekar, P. (1999) 'Brides who travel: gender, transnationalism, and nationalism in Hindi film', *positions* 7: 730–61.

Norton, R. (2000) 'Reconciling ethnicity and nation: contending discourses in Fiji's constitutional reform', *Contemporary Pacific* 12: 83–122.

Ong, A. (1999) *Flexible Citizenship: The Cultural Logics of Transnationality*, Durham and London: Duke University Press.

Oza, R. (2001) 'Showcasing India: gender, geography, and globalization', *Signs* 26(4): 1067–95.

Patel, S. (1988) 'Construction and reconstruction of women in Gandhi', *Economic and Political Weekly*, 20 February: 377–87.

Sengupta, H. (2001) 'Incidents of violence, vandalism mar Valentine's day', *India Abroad*, 23 February: 24.

Srivastava, S. (2001) 'Non-Gandhian sexuality, commodity cultures, and a "happy married life": the cultures of masculinity and heterosexuality in India', *South Asia* 24: 225–49.

Thomas, R. (1995) 'Melodrama and the negotiation of morality in mainstream Hindi film', in Breckenridge, C. (ed.) *Consuming Modernity: Public Culture in a South Asian World*, Minneapolis: University of Minnesota Press.

Thussu, D.K. (2000) 'The Hinglish hegemony: the impact of western television on broadcasting in India', in French, D. and Richards, M. (eds) *Television in Contemporary Asia*, New Delhi: Sage Publications.

Uberoi, P. (1998) 'The Diaspora comes home: disciplining desire in DDLJ', *Contributions to Indian Sociology* 32: 305–36.

Zimmer-Tamakoshi, L. (1993) 'Nationalism and sexuality in Papua New Guinea', *Pacific Studies* 16: 61–97.

Further reading

Chatterjee, P. (1989) 'Colonialism, nationalism, and colonized women', *American Ethnologist* 16: 622–33, remains the best description of the development of women's domesticity as a sign for nation in the Indian nationalist movement.

Derné, S. (2000) *Movies, Masculinity and Modernity*, Westport, Connecticut: Greenwood Publishing Group, provides a good description of contemporary male filmgoers' emphasis on monopolizing public domestic spaces.

Mankekar, P. (1999) 'Brides who travel: gender, transnationalism and nationalism in Hindi film', *Positions* 7: 730–61, provides a good description of recent Hindi films' emphasis on restrictions on women's transnational movement.

8

LATIN AMERICAN URBAN MASCULINITIES
Going beyond 'the macho'

Katie Willis

Summary

Men in Latin America are often portrayed as being violent, aggressive and womanizing *machos*. This chapter considers the lives of urban men throughout the region, and argues that there is a diversity of ways in which masculinity can be performed. Spatially, it is argued that in Latin America there is a distinction between the masculine 'public' sphere of the street, work and politics, and the domestic, feminine 'private' sphere of the home. Transcending these boundaries has implications for constructions of gender identities. This chapter shows the fluidity of the public/private distinction and how masculinity is socially-constructed within particular contexts. Changes in employment, education and social attitudes have increasingly challenged gender norms, leading to more diverse experiences for the region's men and women.

- Work
- Home
- Non-heterosexual spaces.

Introduction

I have to work so hard because I have four women to support. You know what us Mexican men are like!

This comment was made to me by a taxi driver in Puerto Escondido, Mexico, in June 2002. He continued saying that it was particularly difficult, as they all lived in the same house. By this point it was clear that there was more to this story than first appeared and he soon revealed that in fact he was talking about his wife and three daughters. He was clearly entertained that he had 'conned' the two British women in his taxi into believing he was a 'real' Mexican *macho* when in fact he was a

97

hard-working man trying to support his family in difficult times. We spent the rest of the journey discussing the economic problems resulting from the downturn in the tourist industry and how hard it was to make ends meet.

As Gutmann (1996) has shown in great detail, the construction of Latin American, and particularly Mexican men, as *macho* is both dangerous and misleading. Through a *macho* identity, men gain status 'as men' through aggression and violence towards each other, as well as towards female partners and children; drunkenness, womanizing and fathering many children to demonstrate virility (Melhuus 1996; Sternberg 2000; Stevens 1973; Szasz 1998). However, as the taxi driver's story demonstrates, there is more than one way to 'be a man' in Latin American cities. It does not mean that the *macho* does not exist, just that constructions of masculinity vary spatially and temporally.

The economic problems highlighted by our taxi driver, provide an introduction to temporal changes. If a key, if not *the* key role for men in a society, is to be the main household breadwinner, then what happens in periods of economic hardship? Changing social norms due to feminism and women's organizations, or media influences from other countries, also mean that identities are constructed in dynamic environments.

The importance of 'space' in the construction of both male and female identities also contributes to the fluidity of these identities. Within the framework of *machismo*, men are supposed to occupy the 'public' space outside the home; a space imbued with meanings of power in the economic and political spheres. As Fuller, in her discussion of masculinity in urban Peru, states:

> Outside space consists of the public and the street. The street is associated with virility and is a dimension of the outside world that is disorderly and opposed to the domestic realm. It is the arena of competition, rivalry and seduction. From an early age the peer group transmits a masculine culture of the street that is opposed to domesticity and centred on the development of strength and virility. Peer groups transmit to boys one of the most important messages of masculine culture: to be a man signifies breaking some of the rules of the domestic world.
>
> (2003: 137)

In contrast, women are supposedly confined to the private or domestic sphere, where they perform roles of 'good mothers' and 'good wives' as part of what Stevens (1973) termed *marianismo*. As elsewhere in the world, the public/private dualism has never been as rigid at it is often presented, but throughout Latin America, men have much greater participation rates in remunerative employment and domestic chores are predominantly carried out by women (Chant with Craske 2003). Variations in these roles and the spaces occupied can clearly, therefore, affect constructions of masculinity and the acceptability of these constructions to the individuals and to the wider society.

It is not only the public/private spatial divisions which are implicated in the construction of gender identities. Within cities there are particular spaces, such as bars, which are understood as 'male spaces' within which certain forms of 'masculine behaviour' are acceptable and where women are often explicitly excluded. At a larger scale, there are national spaces and constructions of appropriate masculinity within them (see Gutmann 1996 for a discussion of masculinity and constructions of Mexican nationhood).

The diversity of male identities within the region has been recognized by researchers, with important implications both for academic understandings of Latin American masculinities and also policy interventions (see Viveros Vigoya 2003 for a useful overview). Research has highlighted how masculinity is socially-constructed, in the same way as women's roles and identities are context-dependent. In addition, the constructions of appropriate male and female behaviour are inter-related, as what it is 'to be a man' may be partly based on not behaving 'like a woman'. Thus, projects aiming to challenge existing relations of inequality between men and women may need men's involvement, rather than focusing purely on women as the agents of social change. In addition, there has been a recognition that men may have particular problems because of their gender and social expectations placed upon them (Chant and Gutmann 2000, 2002; Jackson 2000).

In this chapter I will focus on the urban spaces of 'work', 'home' and 'non-heterosexual spaces'. I will also focus on low-income groups (sometimes termed the 'popular sector' in Latin America), partly because of word limits, but also due to the relative scarcity of work on middle-class gender relations and identities in the region (although see some chapters in Gutmann 2003; Willis 2000). Throughout the chapter it is important to remember that the 'spaces' discussed are not just physical spaces, but are places resonating with social meanings, in this case, intertwined with notions of appropriate male and female behaviour. In addition, the chapter will highlight that while the focus is on male identities and performances, the roles of women are implicated in the constructions of masculinity (Lancaster 1992; Sternberg 2000).

Work

The spatial separation of 'home' and 'work' is clearly complicated by the development of home-based income-generating activities. This may create opportunities for shifting intra-household gender relations and understandings of masculinity (Pineda 2000 discussed below). In addition, while 'work' activities may take place outside the home, 'doing work' can provide status which also has currency in other spheres (Fuller 2000). As the Puerto Escondido taxi driver stressed, his role, and indeed his status as the main breadwinner within the home, was being threatened by economic problems. As elsewhere in the world, the rootedness of male identity within men's activities as income providers for the family is a key dimension of constructions of masculinity, and if this position is challenged, male feelings of self-esteem may suffer (Fuller 2003). In Latin America, some authors have described this role of provider as 'positive *machismo*', or an alternative construction of the *macho* identity (Mirandé 1997; Varley and Blasco 2000).

Women have always been engaged in paid activities outside the home in Latin America, but since the 1950s women's participation in the labour force has risen as a result of increasing education and in some cases changing attitudes to women's wage earning (Chant with Craske 2003; Psacharopoulos and Tzannotos 1992). In addition, wider economic changes through the expansion of service sector employment, the growth of export-processing factories in some parts of the region, and the widespread impacts of economic crisis and recession on household income have all contributed to higher levels of women's employment (Pearson 2000). As a result, 'spaces of work' in general have become less numerically-dominated by men. This does not mean that *all* sites of work are being feminized, as there are still occupations that are regarded largely as the preserve of men. In addition, even in workplaces where women are numerically dominant, managerial positions, particularly senior key decision-making posts, are often occupied by men (Pearson 1995).

Within a framework of *machismo*, the challenge to men's position as family breadwinner by women's entry into the labour force, particularly if this involved moving into the public world beyond the boundaries of the home, would be regarded as unacceptable. In some cases, there is evidence of men preventing wives from working outside the home, so reasserting their power within the household. This exercise of power also presents an image of control over family members to other people, particularly other men. Sylvia Chant, in her work in Guanacaste, Costa Rica, quotes men discussing their opinions on women working outside the home. It is very clear that men's positions of power are challenged by this, as women gain access to income and opportunities to mix with others, particularly with other men. Chant quotes Don José, who is now separated from his common-law wife after 37 years together. He never allowed her to work outside the home when they were together, because, as he says, when women work they 'become different, want to dress themselves up more, go to dances and have a good time'. In addition, relations within the household change, particularly if women earn more than their husbands: 'women who earn more become boss [in the household] . . . they are freer' (2000: 211).

In other cases, women's entry into the labour force is accepted at one level, but may lead to an attempt to assert power through other means. This can be achieved through increased levels of domestic violence as men feel their superior positions within the household are threatened by women's access to income. In her longitudinal study of households in low-income settlements in Guayaquil, Ecuador, Caroline Moser (1992) describes how levels of domestic violence rose following increased levels of male unemployment and larger numbers of women going into the labour force due to macro-economic restructuring in the late 1980s. With their position as provider perceived as undermined, men's frustrations were taken out partly on family members and displays of violence and aggression were a way of reasserting authority within the home. Other strategies adopted by men in the face of such shifts in the domestic division of labour included abandonment of the family, contributing to rising numbers of female-headed households (see also Chant 2000).

Of course, for many men, having a female partner in the labour force does not necessarily lead to a violent reaction. It is clear, however, that for many Latin American men, as in many other parts of the world, the undermining of the breadwinner role, particularly if female partners become the main provider, is wounding. Luís, another Costa Rican man interviewed by Chant, said that if a man cannot provide for his wife and children, 'his self-image and his image in the eyes of others "*ya no vale nada*" [isn't worth anything]' (2000: 211).

José Olavarría (2003), in his study of Chilean working-class fathers in Santiago, Chile, found similar patterns and quotes Alexis, a 39-year-old husband and father:

> I have never liked having my wife work; I want to have a job myself and have her take care of the children until they grow up. I work and I provide for all their needs. I fulfil all my duties, [and] I pay the bills. Maybe this is a very chauvinistic system, but I feel self-sufficient because I was raised this way. I was raised to look out for my family.
>
> (2003: 336)

Olavarría discusses how such discourses reflect the awareness many Chilean men have of the changing material context in which they live, where women are increasingly engaged in paid employment. In addition, Alexis' quote demonstrates a recognition that having these attitudes could be regarded (by the researcher at least) as undesirable or what he terms 'chauvinistic'. Fuller (2003) describes a similar pattern in Peru, where discourses of gender equality, especially within marriage, clash with continued associations of male power and privilege through the breadwinner role.

Home

'Home' as a social space laden with gendered power relations, has been a key site in the construction of masculinities and femininities in Latin America. Within the home, although the 'domestic' is regarded as 'feminine', men are supposedly dominant through both their income-generating capacity, as described above, and also their decision-making roles. In some cases this power can be reinforced or expressed through domestic violence. The construction of masculinity within the home is counterposed against women's behaviour and particularly the role of wives, mothers and daughters in domestic chores.

However, this division of labour and construction of power relations is not ubiquitous, particularly in periods of great economic change. Two contrasting examples from my research in Oaxaca City, southern Mexico, can be used to highlight diverse performances of masculinity within the 'homespace'. Alejandro and César (both pseudonyms) are in their fifties, work in informal-sector occupations, and are married with children. Intra-household gender relations are, however, very

Figure 8.1 Masculinity within the 'homespace' (*source:* Katie Willis).

different, and their performances of masculinity also vary. Alejandro and his wife, Maria, run a small grocery shop in one of the informal settlements overlooking the city. In addition, he does tailoring work for neighbours, altering shop-bought clothes, or sewing made-to-measure garments. Maria tends to go to the market to purchase food and shop supplies, while Alejandro stays at home and minds the shop. He also helps his daughter-in-law look after his baby grandson. They live, along with Alejandro's son, in a room in the house having married a couple of years ago.

In contrast, César works as a carpenter in a workshop in the same settlement. Despite the fact that the workshop adjoins the corrugated iron house where he lives with his wife and five children (aged 8 to 17), he rarely comes into the house except to eat, and is often seen in a drunken state in the street and in the city centre. His wife, Graciela, works long hours as a domestic servant, and the children also do errands for neighbours to earn a few pesos.

Violence is common within the household, with César beating his wife and children, and his sons policing their sister's behaviour by beating her if she goes out with her boyfriend alone. Gossip in the community was that César had a mistress in the city centre, but I never found out if this was true.

Alejandro and César view the homespace differently, with one conforming more to a *macho* identity, explicitly exerting power through his violence and his non-performance of domestic tasks. Within the wider community, many residents talked disparagingly about César, his drunkenness and his perceived neglect of his family, while Alejandro seemed to be well-regarded and had been elected to the local residents' committee.

For both men, however, domesticity has its limits. For César, these were clear, but for Alejandro while childcare and joint decision-making regarding household expenditure and budgeting were acceptable, other domestic chores such as washing and cooking were carried out by his wife and daughter-in-law. These patterns have been described by Gutmann (1996) in his work in Colonia Santa Domingo, Mexico City and in Mexican national surveys (see Table 8.1). Gutmann also highlights how it is not only men who do not want to participate in these chores, but that women are wary of men who are so domesticated, feeling that they lack 'masculinity' (see also Olavarría 2003: 341).

As Varley and Blasco (2000: 117–18) argue, increasing research on masculinity within Latin America has unsettled the static notion of *machismo* and the *macho*, resulting in a 'bewildering variety of meanings of masculinity'. However, there is less research examining changing performances of masculinity over time (although see Gutmann 1996). The shifts in Latin American urban economies may provide some indications of temporal shifts. For example, Pineda's Colombian study of men working in their wives' home-based businesses demonstrates that 'working at home is potentially, but not necessarily, a route to men's everyday participation in reproductive work' (2000: 79). Because of increasing levels of male unemployment in

Table 8.1 Percentage of population aged 20 and over who carry out domestic tasks, by gender (Mexico, 1996)

Task	Women (%)	Men (%)
Cleaning the house	85.6	20.4
Food preparation	85.1	12.4
Clothes washing	84.0	6.7
Washing dishes	82.8	9.5
Ironing	71.0	6.9
Childcare	48.7	23.2
Caring for the sick	4.0	1.6
Caring for the elderly	2.7	0.9

Source: INEGI (2002).

Note
These figures are based on a survey in both urban and rural areas of Mexico.

Cali, Colombia, men are working in the micro-enterprises set up by their female partners with funds from micro-credit schemes. For example, Pineda quotes Edgar who 'helps' his female partner:

> As it is the economic situation in the country, where many companies all over the place are going bankrupt or are laying off staff, then what happens is, many men become unemployed. At this moment it is easier to get a job for a woman than a man . . . for a man it's very difficult, it's here where a woman comes to occupy the place of the man and a man must perform in place of a woman.
>
> (ibid.)

This quote encapsulates the idea of spaces occupied by men and women shifting because of macro-economic change. However, it was only in some cases that this spatial shift was associated with changing performances on the part of men as they became involved in daily reproductive tasks. This involvement could be for a range of reasons including their lack of bargaining power linked to women's self-empowerment due to their business (see also Olavarría 2003). In addition, Pineda argues that the 'invisibility' of housework decreases when men and women are both working at home together, and some of his male respondents said this made it easier to share the domestic tasks. Men's greater involvement in domestic chores was not, however, inherent within the process of women's entry into self-employment.

The massive expansion in women's multinational corporations (MNC) factory employment in some parts of the region has also created the potential for shifting gender roles and relations within the home. Safa (1995) in her work in Puerto Rico, Cuba and the Dominican Republic highlights some households where women's factory employment has been associated with greater participation in housework by their male partners. However, this is not automatic and tends to be when partners have higher levels of education and have not recently migrated from rural areas. Cravey (1997) also discusses men's greater involvement in domestic chores with women's entry in factory work, but also points out that, as with Alejandro and César in Oaxaca, certain household tasks, especially laundry, are very rarely carried out regularly by men (see also Table 8.1). She also describes a man who, while regularly involved in childcare, does not freely admit to this 'as if it contradicted his self-image' (176). Thus, while the ways of 'being a man' are certainly less restrictive than suggested by the *macho* stereotype, the importance of paid work to a man's self-image and also the degree of involvement in reproductive activities remain key.

Non-heterosexual spaces

As in the North, in Latin America the spaces constructed as 'non-heterosexual' tend to be concentrated in urban areas, and within these cities and towns there are certain places which are recognized as arenas within which, at least at certain times of the day, non-heterosexual masculine identities can be performed (Parker 1999). This may be at particular bars and clubs, or cruising areas in parks or on the street for example. Before discussing particular groups and spaces, it is important to highlight some of the complexities associated with identity and sexual activity between men. While I feel rather uncomfortable with the notion of 'non-heterosexual spaces' because it implies a 'norm' of heterosexuality, calling such spaces 'gay spaces' or 'queer spaces' would fail to recognize the myriad of ways in which men engaging in sex with other men self-identify or are identified by other members of society.

Ethnographic research in many Latin American countries has stressed the fact that with sexual activity between men, social stigma and the identification as 'homosexual' are often only assigned to the man who is penetrated (the 'passive' role) (Buffington 1997; Lancaster 1992; Lumsden 1996; Parker 1999). Thus, as with constructions of the *macho* outlined at the start of the paper, sexual prowess in terms of 'active' sexuality, is regarded 'positively'. Michael Higgins and Tanya Coen (2000), in their work with a group of male transvestite prostitutes in Oaxaca City, Mexico detail the strong heterosexual identity of the prostitutes' boyfriends. These men do not identify as 'gay' or 'homosexual', rather they see themselves as 'real men', taking the 'active' role in sex and protecting their 'girlfriends' who dress in overtly 'feminine' ways with tight dresses, high heels, false nails and large amounts of make-up.

For many men (especially those on low incomes) who are sexually attracted to men, the opportunities for expressing this identity are often restricted to certain forms of performance. It may be in the form of paid sex work, or as a client in this relationship, in which case particular urban spaces will be the sites of exchange. Another space for the expression of this identity could be shared accommodation with similar men, often using income from sex work to finance day-to-day living (Kulick 1998; Prieur 1998; Schifter 1998). Within these spaces, while there is the freedom to be sexual towards men, particular forms of male performance seem to be encouraged. A number of ethnographic studies highlight the importance of what could be termed 'camp behaviour' and transvestism within these sites, where men who do not fit into such roles are excluded.

Sam Quinones, in his fascinating book of vignettes on Mexican life, quotes Juan Carlos Hernández, an anthropologist:

> When I was a boy and I began to realize I was gay, I began to say to myself 'I don't want to dress like a woman'. But in my family there was no way of being gay other than dressing like a woman. I think this happens to a lot of gays in Mexico. There's no masculine model of gay life, one that doesn't involve cross-dressing. Mexicans don't tolerate, say, a masculine soccer player who also says, 'Yes, I like men'.
>
> (Quinones 2001: 80)

For Hernández, the experience of being gay in Mexico was complicated by the fact that the way he wanted to express his sexuality conflicted with his family's and wider society's constructions of male homosexuals. Dressing and behaving 'like women' was regarded as the only way gay men could express their sexual identities. This does not mean that this form of performance is regarded as acceptable by all, as homophobia is present throughout the region regardless of how same-sex desire is presented (Chant with Craske 2003; Green and Babb 2002; NACLA 1998).

The previously described, somewhat limiting forms of 'non-heterosexual performance' are being challenged, as Parker (2003) and Carillo (2003) show in relation to Brazil and Mexico respectively. It is increasingly possible to talk about a 'gay' identity within which there are multiple forms of expression. Such changes, they argue, are both a reflection of, and a response to, the increase in commercial sites, such as clubs and bars. In addition, political mobilization and activism around HIV/AIDS have also provided a space within which men with same-sex desires can explore their identities beyond the stereotypes of the *macho* or the effeminate transvestite.

Conclusions

This chapter has provided a brief overview of some of the ways in which Latin American masculinities have been constructed. It is clear that while *macho* as a form of male performance and identity is present in the region's cities, it is far from the norm. Within urban areas, the construction of different spaces enables the performance of a range of male identities and these performances in turn shape the nature of urban space. Identities are not fixed, and men's performances *as men* can vary across urban space; a man's behaviour in a bar may be very different from his actions when he is at home playing with his children. Many films and television programmes may continue to play on the stereotype of Latin American men as violent, hard-drinking, womanizers, but the reality is incredibly complex.

In policy terms, increased awareness of the variations in male performance and the socially-constructed nature of masculinities has led to new spaces opening up for men within the non-governmental organization (NGO) world. For example, the Centre for Communication and Popular Education (CANTERA) (Welsh 2001) and the Group of Men Against Violence (Broadbent 2001) run workshops in Managua, Nicaragua. These provide men with spaces to discuss their problems, share experiences and consider how their roles are affected by social pressures (see also Sternberg 2000). While these new political spaces are relatively rare, they do highlight the constructed nature of masculinity and how in some cases, NGO activities can help men to analyse and change behaviour that they feel is destructive to themselves and to their families.

Latin American cities provide a multiplicity of spaces within which different forms of masculinity can be developed and performed. However, despite these opportunities, it is clear that men's abilities and desire to express themselves in different ways are constrained by wider social norms, as well as economic, political and cultural processes. The range of 'acceptable' masculinities is certainly wider than the stereotypical *macho*, but given existing, albeit dynamic, systems of patriarchy, for most men, there remain boundaries to the performance of male identities.

References

Broadbent, L. (2001) *Macho*, Video. Glasgow: Broadbent Productions.

Buffington, R. (1997) 'Los Jotos: contested visions of homosexuality in modern Mexico', in Balderston, D. and Guy, D. (eds) *Sex and Sexuality in Latin America*, New York: New York University Press.

Carillo, H. (2003) 'Neither *machos* nor *maricones*: masculinity and emerging male homosexual identities in Mexico', in Gutmann, M. (ed.) *Changing Men and Masculinities in Latin America*, London: Duke University Press.

Chant, S. (2000) 'Men in crisis? Reflections on masculinities, work and family in north-west Costa Rica', *European Journal of Development Research* 12(2): 199–218.

Chant, S. with Craske, N. (2003) *Gender in Latin America*, London: Latin American Bureau.

Chant, S. and Gutmann, M. (2000) *Mainstreaming Men into Gender and Development: Debates, Reflections and Experiences*, Oxford: Oxfam.

Chant, S. and Gutmann, M. (2002) '"Men-streaming" gender? Questions for gender and development policy in the twenty-first century', *Progress in Development Studies* 2: 269–82.

Cravey, A. (1997) 'The politics of reproduction: households in the Mexican industrial transition', *Economic Geography* 73: 166–86.

Fuller, N. (2000) 'Work and masculinity among Peruvian urban men', *European Journal of Development Research* 12(2): 93–114.

Fuller, N. (2003) 'The social construction of gender identity among Peruvian males', in Gutmann, M. (ed.) *Changing Men and Masculinities in Latin America*, London: Duke University Press.

Green, J.N. and Babb, F.E. (2002) 'Introduction to special issue on gender, sexuality and same-sex desire in Latin America', *Latin American Perspectives* 29(2): 3–23.

Gutmann, M. (1996) *The Meanings of Macho: Being a Man in Mexico City*, Berkeley: University of California Press.

Gutmann, M. (ed.) (2003) *Changing Men and Masculinities in Latin America*, London: Duke University Press.

Higgins, M.J. and Coen, T.L. (2000) *Streets, Bedrooms and Patios: The Ordinariness of Diversity in Urban Oaxaca*, Austin: University of Texas Press.

INEGI (2002) *Mujeres y Hombres 2002*, Aguascalientes: INEGI.

Jackson, C. (2000) 'Men at work', *European Journal of Development Research* 12(2): 1–22.

Kulick, D. (1998) *Travesti: Sex, Gender and Culture Among Brazilian Transgendered Prostitutes*, Chicago: University of Chicago Press.

Lancaster, R. (1992) *Life is Hard: Machismo, Danger and the Intimacy of Power in Nicaragua*, Berkeley: University of California Press.

Lumsden, I. (1996) *Machos, Maricones and Gays: Cuba and Homosexuality*, Philadelphia: Temple University Press.

Melhuus, M. (1996) 'Power, value and the ambiguous meanings of gender', in Melhuus, M. and Stølen, K.A. (eds) *Machos, Mistresses and Madonnas*, London: Verso.

Mirandé, A. (1997) *Hombres y Machos: Masculinity and Latino Culture*, Oxford: Westview.

Moser, C. (1992) 'Adjustment from below: low-income women, time and the triple role in Guayaquil, Ecuador', in Afshar, H. and Dennis, C. (eds) *Women and Adjustment Policies in the Third World*, London: Macmillan.

NACLA (1998) 'Sexual politics in Latin America', *NACLA Report on the Americas* XXXI(4): 16–44.

Olavarría, J. (2003) 'Men at home?: childrearing and housekeeping among Chilean working-class fathers', in Gutmann, M. (ed.) *Changing Men and Masculinities in Latin America*, London: Duke University Press.

Parker, R. (1999) *Beneath the Equator: Cultures of Desire, Male Homosexuality and Emerging Gay Communities in Brazil*, London: Routledge.

Parker, R (2003) 'Changing sexualities: masculinity and male homosexuality in Brazil', in Gutmann, M. (ed.) *Changing Men and Masculinities in Latin America*, London: Duke University Press.

Pearson, R. (1995). 'Male bias and women's work in Mexico's border industries', in Elson, D. (ed.) *Male Bias in the Development Process*, Manchester: Manchester University Press.

Pearson, R. (2000) 'All change? Men, women and reproductive work in the global economy', *European Journal of Development Research* 12(2): 219–37.

Pineda, J. (2000) 'Partners in women-headed households: emerging masculinities?', *European Journal of Development Research* 12(2): 72–92.

Prieur, A. (1998) *Mema's House: Mexico City*, Chicago: University of Chicago Press.

Psacharopoulos, G. and Tzannatos, Z. (1992) *Women's Employment and Pay in Latin America: Overview and Methodology*, Washington, DC: World Bank.

Quinones, S. (2001) *True Tales from Another Mexico: The Lynch Mob, the Popsicle Kings, Chalino and the Bronx*, Albuquerque: University of New Mexico Press.

Safa, H. (1995) *The Myth of the Male Breadwinner*, Oxford: Westview.

Schifter, J. (1998) *Lila's House: Male Prostitution in Latin America*, London: Harrington Park Press.

Sternberg, P. (2000) 'Challenging machismo: promoting sexual and reproductive health with Nicaraguan men', *Gender and Development* 8(1): 89–99.

Stevens, E. (1973) '*Marianismo*: the other face of *machismo* in Latin America', in Pescatello, A. (ed.) *Male and Female in Latin America*, Pittsburgh: University of Pittsburgh Press.

Szasz, I. (1998) 'Masculine identity and the meanings of sexuality', *Reproductive Health Matters* 6(12): 96–104.

Varley, A. and Blasco, M. (2000) 'Exiled to the home: masculinity and ageing in urban Mexico', *European Journal of Development Research* 12(2): 115–38.

Viveros Vigoya, M. (2003) 'Contemporary Latin American perspectives on masculinity' in Gutmann, M. (ed.) *Changing Men and Masculinities in Latin America*, London: Duke University Press.

Welsh, P. (2001) *Men aren't from Mars: Unlearning Machismo from Nicaragua*, London: CIIR.

Willis, K. (2000) 'No es fácil, pero es posible: the maintenance of middle-class women-headed households in Mexico', *European Review of Latin American and Caribbean Studies* 69: 29–45.

Further reading

Chant, S. with Craske, N. (2003) *Gender in Latin America*, London: Latin American Bureau. An excellent introduction to gender debates in Latin America from which to develop an understanding of men's identities in the region. While the focus is largely on women, there are a number of sections on men and masculinities.

Gutmann, M. (ed.) (2003) *Changing Men and Masculinities in Latin America*, London: Duke University Press. A very useful collection of chapters on the diversity of masculinities within Latin America. The first two chapters provide very useful overviews of key themes in the field, and the remaining 14 chapters focus on different countries or aspects of men's lives. Topics covered include violence, fatherhood, sexuality and alcoholism.

Parker, R. (1999) *Beneath the Equator: Cultures of Desire, Male Homosexuality and Emerging Gay Communities in Brazil*, London: Routledge. Fascinating account of male homosexuality in urban Brazil, with a particular focus on spatial practices.

Welsh, P. (2001) *Men Aren't from Mars: Unlearning Machismo from Nicaragua*, London: CIIR. A short publication which provides excellent material about how NGOs have begun to work with men to examine the social constructions of masculinity and the effects of such constructions. Written by a long-term NGO activist in the region.

Part 3

MASCULINITY AND VIOLENCE

9

ETHNIC VIOLENCE AND CRISES OF MASCULINITY

Lebanon in comparative perspective

Michael Johnson

Summary

This chapter examines the connections between ethnic violence, rural-urban migration, changing family structures and masculinities in the case of Lebanon. It argues that urbanization and the emergence of nuclear family forms have challenged traditional masculine identities. The resulting tensions have been partly resolved by adopting stronger ethnic identities. Ethnic nationalism is shown to have filled the moral vacuum created by the break from the ordered world of rural extended-kinship structures, and to have helped resolve difficult tensions within the urban nuclear family.

- Ethnicity and the city
- Maronite Catholics and Shia Muslims
- Ethnicity and kinship
- Ethnicity and masculinity
- Ethnicity and male fighters
- Honour and masculinity.

Introduction

Wherever they are fought – in Lebanon (Johnson 2001), Bosnia (Gutman 1993) or India (Das 1997; Menon and Bhasin 1998) – what is striking about ethnic or communal wars is that the style of violence is so gruesome and nasty. The aim, it seems, is not just to 'cleanse' the space of the homeland by driving out or killing the Other, but to kill in particularly brutal ways, involving the violation and mutilation of the bodies of men and women, and thus the utter degradation and humiliation of the enemy community.

The mutilations and other acts of war are heavy with symbolism, involving a particular construction of masculinity associated with the cultural values of honour and shame. Male bodies are castrated to strike at their manhood and reproductive capacity. Women are raped not just to provide recreation for the fighters, but also to shame the victims' men who should have protected them. The wombs of pregnant women are cut open to destroy the accursed progeny of the hated Other. Churches, mosques and temples are destroyed or burnt down; graveyards are dug up with bulldozers; everything that represents the Other – including the Other's space – has to be destroyed, erased, obliterated.

Fought between 1975 and the 1990s, the Lebanese civil wars were immensely complicated, but for much of the time they involved fighting between a variety of Christian and Muslim confessional communities, particularly Maronite and Greek Catholics, and Sunni, Shia and Druze Muslims. As members of these groups did not usually intermarry, they tended to see themselves as belonging to distinct ethnic groups defined by exclusive lines of descent.

Many if not most of the fighters were recruited from the Maronite Catholic and Shia Muslim communities, which had migrated to the city of Beirut relatively recently. The fighters were members of militia-based parties whose political programmes appealed to men whose masculine identities were threatened by urban living. This essay explores these issues in a comparative context, and argues that ethnic warfare is closely associated with contradictions that develop within the urban nuclear family.

Ethnicity and the city

Strong ethnic and communal identities are, like nationalism (Anderson 1983; Gellner 1983; Hobsbawm 1990), modern phenomena. Modernity undermines an established order, and this can lead to the emergence of conflict in multi-ethnic societies that previously had been relatively harmonious (Fox 1997). Thus, in Lebanon, confessional conflict between Muslims and Christians first broke out in the nineteenth century, when the agrarian or 'feudal' order was undermined by capitalism (Johnson 2001). But while such a relationship can be documented, it is difficult to specify and rank the elements of modern society that contribute to ethnic conflict.

One change that seems especially influential is the shift from extended to nuclear families. Modernity undermines the extended family and threatens patriarchal authority, and a number of writers have seen an important relationship between this process and the development of extreme forms of ethnic identification (Fromm 1955; Eriksen 1993; Brown 1994; Johnson 2001). Often the breakdown of extended kinship structures is associated with migration to the city, and early anthropological studies in the Copperbelt of Northern Rhodesia (Mitchell 1956; Epstein 1958) through to more recent accounts of ethnic or communal conflict in the Indian subcontinent (Das 1992; Ludden 1997; Vanaik 1997) all illustrate the way social developments in an urban space can promote ethnic nationalism.

The Copperbelt studies, for example, showed how migrant workers from a countryside where social relations had little to do with ethnicity became far more conscious of their tribal identity in the new mining towns. Cut off from their families who often remained in the villages, many lived in all-male hostels. They faced difficulties in finding housing for their families and in adjusting to the disciplines of wage labour, and it seems that many of them dealt with the alienation of urban life by forming social relationships in venues like clubs and beerhalls with members of their ethnic group or tribe. Similarly in India, recent migrants to the city tend to socialize along communal lines as Hindus or Muslims, and under particular conditions they can become involved in ethnic violence.

The move to the city tends to break the link between men and their rural extended families, and the ethnic group can provide a 'substitute family' that helps recent migrants make the transition to urban life, not just in the material provision of housing, jobs and other patronage services, but also in the promotion of emotional security. When families are re-established in the city, they tend to take a nuclear form. The patriarchal head of the urban family then finds himself in a difficult and emotionally uncertain role, responsible as the family's provider and protector of its public image. He has to operate in a competitive economic environment, and he fears the consequences of failing to provide a sufficient income to support his wife and children. He also worries that his family might become morally corrupted by the city, and he imposes a strict repression on his wife and children to control their sexuality and general behaviour.

This repression is in some respects similar to that which existed in the rural extended family, but in the city there are more opportunities for filial and marital rebellion, and the responsibility for patriarchal control falls more heavily on the shoulders of the paterfamilias instead of being shared among the elders of the clan. This, then, is a new form of patriarchy, related to that which Hisham Sharabi (1988) identifies in the modern Middle East and calls 'neopatriarchy'.

As we shall see, ethnic nationalism helps resolve the contradictions within the repressive family by defining roles: first, for patriarchs who control the family as the ethnic leader dominates the 'nation'; second, for women who are elevated as 'mothers of the nation' or the 'mothers of martyrs'; and third, for young men who become heroic fighters for the nation's honour (Johnson 2001: 200–2).

Maronite Catholics and Shia Muslims

Maronite Christians (belonging to the Lebanese branch of the Catholic church) migrated to Beirut in large numbers during the 1950s. Land shortage on the intensively farmed mountains, and expanding opportunities in Beirut's trading and banking sectors, led to a dramatic growth of the Christian community in the city. But while these new migrants usually found work and housing, many felt excluded from the patronage networks in the old-established Christian quarters, and were attracted to a 'neo-fascist' party – known in French as the Phalanges – which represented a predominantly lower middle-class and Maronite interest. Significantly, this militia-based party also laid great emphasis on linking confessional, nationalist and kinship values as expressed, for example, in its motto: 'God, father-land and family' (Entelis 1974).

After the 1950s, the next wave of immigration to Beirut was predominantly Shia Muslim. During the 1960s and 1970s, thousands of Shia peasants were forced to leave agriculture (Nasr 1985; Sayigh 1994). In the shanty towns of the city, Shia from southern Lebanon met their co-religionists from the north, and for perhaps the first time both groups began to 'imagine' (Anderson 1983) themselves as part of the same community. This new identity was encouraged by some members of a Shia intelligentsia who had benefited from the expansion of state education during the 1960s. Such people as lower-level civil servants and teachers in government schools provided a new intellectual leadership that had been lacking in an under-schooled agrarian society.

The leader of this popular Shia mobilization was the head of the religious hier-archy: the Imam Musa as-Sadr (Ajami 1986). In March 1974 he organized a mass rally, and close to 100,000 Shia from all parts of Lebanon attended, many of them carrying Kalashnikov automatic rifles, and some of them rocket launchers. In a speech punctuated by a militant and jubilant firing of guns into the air, the Imam told the crowd that 'arms are an ornament of men' (Salibi 1976: 78), and he called on his followers to rise up against the tyranny that oppressed them even if it meant paying with their blood as martyrs to the cause. Later he formed his own militia called Amal (or 'hope').

In his studies of Maronite and Shia migrants in Beirut, Fuad Khuri (1972, 1975) shows how confessional identity was of little or no significance in the village. Family and clan allegiance was much more central in rural social networks, patronage-dis-tribution and politics. It seems that it was only when the rural extended family was undermined by migration to the more individualistic society of the city that confes-sionalism and recruitment to communal parties developed to a significant extent. The Christian Phalanges and Muslim Amal did recruit in the countryside, especially as the Lebanese civil wars progressed; but they were initially urban movements and appealed mainly to relatively recent migrants to Beirut who found it difficult to find patrons to assist them in finding jobs, housing and education.

What made the Phalanges and Amal so successful, however, was their ability to appeal to people's values as well as their material needs (Johnson 2001). Their pro-grammes of ethnic nationalism filled the moral vacuum created by the break from the ordered world of rural extended-kinship structures, and they also helped to resolve what were difficult tensions within the urban nuclear family.

Ethnicity and kinship

The widespread adoption of ethnic identities as a partial substitute for kinship organization contributes to David Brown's discussion of ethnicity as an 'analogy' of the family (1994: xviii and 17–19) and Thomas Eriksen's of nationalism as 'metaphoric kinship' (1993: 107–8). What these and other authors are getting at is not just the way ethnic nationalists talk about homelands as 'motherlands' or leaders as 'fathers', but the way they often identify their ethnic community as ascriptive and bound together by common descent. Such communal groups as Maronites and Shia in Lebanon, or Hindus and Muslims in India, are not absolutely endogamous but they are so as an ideal.

In Lebanon, marriages between Maronite and Greek Catholics are acceptable, between Catholics and Eastern Orthodox Christians less so, and between Christians and Muslims virtually anathema (Hanf 1993; Khlat 1989). On the rare occasions when Christian-Muslim marriages occur outside the small liberal elite, the offending couple are often ostracized by their respective communities, and young women are sometimes killed by their male relatives to prevent such an abhorrent union.

In many different parts of the world, the ethnic community is 'imagined' and interpreted as a replication and, in some sense, an actual extension of the family via such other communities as clan, caste, tribe and, very significantly, sect or confession. Even where other aspects of the ethnic group such as language or a distinctive national history are perhaps more significant, religious confession is often important as well. Thus Sinhala and Tamils in Sri Lanka are differentiated by their respective allegiances to Buddhism and Hinduism as well as language, and Greeks and Turks in Cyprus by Christianity and Islam.

Insofar as these confessions are endogamous communities, then family relations can easily be imagined as extending into the religiously defined ethnic group. Also, the colonial or imperial experience of these groups often established the principle that personal or family law governing marriage, divorce and inheritance should be based on religious precepts. In India, the British codified Hindu, Muslim and Buddhist personal law, and in the Middle East the Ottomans left such matters to the separate legal systems of the Muslim religious judges, and the courts of the various Christian and Jewish communities.

Such systems of religious law have usually survived in one form or another to the present day. The influence of secular ideas has meant that some aspects of personal status have been reformed. In the Middle East and Indian subcontinent, for example, laws limiting polygyny and the Muslim husband's right to immediate divorce by repudiation were introduced in many countries during the 1950s and 1960s. Nevertheless, laws governing key aspects of family life remain largely religious and help to reinforce the already powerful links between the confessional ethnic community and the kinship groups of which it is imaginatively composed.

Ethnicity and masculinity

Family values are so powerful that social and political threats to them often provoke a romantic reaction. Amrita Chhachhi (1989), in her article on religious 'fundamentalism' in south Asia, provides a convincing argument that Hindu and Muslim revivalist movements in India and Pakistan during the 1980s represented a patriarchal reaction to the improved status of women that had been a consequence of their increased participation in the formal economy and politics. Similarly, the growing support for Amal in Lebanon can be seen, in part, as a romantic reaction by Shia men whose masculinity was threatened by difficulties in finding employment and by the ideas of gender-equality propagated by Lebanese socialist and communist parties.

In Lebanon, the sociopolitical subordination of women was most marked in the Shia 'fundamentalist' Hizbullah (the Party of God). This militia had been founded with the encouragement of Iranian 'revolutionary guards' in the early 1980s. Financed by Iran, Hizbullah provided extensive educational, health and welfare services in the regions under its control, and it competed with Amal for the support of the Shia in Beirut (Saad-Ghorayeb 2001). It was heavily influenced by conservative Shia clerics and, although it later moderated its theocratic ideas, at the start it was committed to the establishment of an Islamic state in at least part of Lebanon. Perhaps inevitably, Hizbullah's attitude to women involved notions of their subordinate status and the need for their seclusion, modest dress, and the covering of their hair with the *hijab* or headscarf.

In the contemporary world, those countries afflicted by ethnic nationalism all seem to share sexual or gender repression. This extends from perhaps rather subtle forms of patriarchy in Northern Ireland and the former Yugoslavia to arranged marriages and the seclusion of women in the Middle East, and even to the revival in north-western India of the ideals of *sati* or widow burning (Chhachhi 1989, 1991). In most of these societies, there is a pronounced sense of male honour and women's shame where it is incumbent on the male to prevent his women descending into a shameless state of being. In Lebanon, 'honour crimes' involving a man killing his daughter, wife or sister for sexual misconduct are often only punished by a token prison sentence of a few months, and the newspapers regularly carry reports of men killing 'their' women in order to protect family honour (Younes 1999).

There is agreement in much of the literature that there exists in the Middle East a widespread belief in the destructive potential of female sexuality (Hibri 1982; Ghoussoub 1987; Accad 1991). Hence the need for men to mount constant surveillance to enforce women's modesty. Women's bodies are seen as being saturated with sexuality. The threat is not just to the woman's proper reproductive role within the family but also – through her ability to seduce healthy and 'normal' men – to the very structure of society and political order. As Kanan Makiya puts it in his devastating study of 'cruelty and silence' in the Arab world, 'Women's bodies are deemed simultaneously the font from which all honor derives and a source of *fitna*, or public sedition' (Makiya 1993: 298).

Of course, this is neither confined to the Middle East, nor to Islamic societies. Similar male attitudes toward women are found among peoples as different as the martial Rajput caste in north-west India (Chhachhi 1989, 1991) and the inhabitants of South African townships (Campbell 1992). Nor are these attitudes a cultural residue from so-called 'traditional' societies. Such patriarchal values are better seen as a reproduction in a new form – as a 'neopatriarchy' – in the 'modern' and urbanized world (Sharabi 1988; Johnson 2001).

Patriarchal authority might even seem to be more pronounced in the city than in the countryside. The veiling and seclusion of women, for example, tend to be urban rather than rural phenomena, partly because they are impractical in the countryside where women work in the fields, and because they are a mark of higher status. But the increased surveillance of female and adolescent members of the family in the city occurs precisely because the urban environment is so threatening as compared with the village. When women shop in the souk or bazaar they might meet other men, and social surveillance is much less efficient in the anonymous city than in the close-knit village community.

The urban patriarchal family is continually undermined by the suppression that sustains it. Husbands fear that their necessary but intrusive surveillance of their wives will drive their women into the arms of another. Also, of course, patriarchs might be seduced by those women who adopt a freer lifestyle in the city, and that is dangerous too. Fathers seek to bring up submissive children but, at the same time, they create the conditions for their sexual and other forms of youthful rebellion.

In this context, ethnic nationalism appears to be an effective response to a threat-ened sense of masculine identity. The imagined community of the 'nation' or ethnic group has an authoritarian and even cruel leader who provides a model for patri-archs. The community also creates a highly valued maternal role for the mothers of the nation, which helps at least some women accept their status within the repres-sive family and thus buttresses patriarchal authority. Finally, and very significantly, ethnic nationalism provides a normatively sanctioned outlet for the frustrations and aggression of sexually repressed young men.

Ethnicity and male fighters

In some ways, the problems presented to young men in the repressive family are more difficult than those faced by women. Adolescent males are encouraged to conform to the values of patriarchy. This presumably explains why they are so disci-plinarian with 'their' women, policing the behaviour of their sisters with beatings and sometimes murder. But they are driven by youthful rebellion as well, and they resent the father's authority to which they feel they should submit. In addition, urbanization and other aspects of modernity undermine their sense of a clearly defined role. They need to establish themselves in a career before they can marry. If there is no family business for them to join, they have to enter a highly competitive market. They are therefore subject to many of the anxieties of their fathers which, especially if they are cast adrift without the emotional security of a wife and chil-dren, are often experienced in a more immediate and frightening form.

The party and militia organizations of ethnic nationalism help by providing an outlet for youthful male rebellion while, at the same time, they promote the links between God, fatherland and family. Young men are encouraged to see them-selves as belonging to a community that replicates the family but relieves them of its contradictions. Whether they are bachelors on their own in the alien city or the sons of repressive urban fathers, ethnicity provides them with a strong sense of self and of their place within a wider society. They can submit to the authority of their ethnic organization and its leader, and because they are part of a higher cause this submission is so much easier than the one forced on them by the tyranny of their fathers.

Ethnic parties and militias give the alienated adolescent an identity, and provide him with a sense of power and control. Such feelings are then enhanced in fighting the Other – as can be illustrated by the testimonies of the fighters themselves (Malarkey 1988; also see Fernea 1985). One Lebanese militiaman said, in 1979, 'I think I liquidated in one day all my problems of identity. At the very moment I got behind the barricade, I became perfectly integrated, totally together.' Another told how the fighting meant 'taking one's future into one's own hands and forgetting family conventions'.

In addition, the discourse and actions of ethnic conflict are replete with sexual imagery, fantasy and aggression, and these are associated with an idealization of the mother and all the pure women of one's own community as compared with the whores of the Other. In 1975 an exchange was overheard on short-wave radio between a Muslim and a Christian, who each accused the other's mother of all manner of wicked depravities. After a particularly invective diatribe, the Muslim said to his adversary, 'You have nothing to say to all that, you son of a whore?' 'No,' replied the Christian, 'I'm much too busy with your sister' (Tabbara 1979: 42).

A short step from this verbal shaming is the humiliation of the Other's honour by raping and killing 'his' women. Of course, some fighters in Lebanon felt ashamed after the event. Some resorted to alcohol or tranquillizers, to hashish or crack cocaine. Others experienced post-traumatic stress disorders, induced by the general horrors of war (Malarkey 1988; Abdennur 1980). It is not necessarily the case that ethnic conflict provides a constant sense of wellbeing, and one fighter in 1979 said, 'When I think about it, I see myself vomiting gobs. I felt this incomprehensible need to go confess myself afterwards. But I couldn't tell the priest anything, nothing' (Malarkey 1988: 304).

In the heat of battle, however, many fighters experience elation and a sense of purpose and integration. The fighting is cathartic and everything permissible: from raping girls to castrating rival men. As a militiaman said in 1984, 'War is my only friend. It's like my wife, I love it. In peace I feel afraid' (Malarkey 1988: 291).

By shaming the Other in this way, it seems the fighters seek to cope with a threatened masculinity that is central to the experience of modernity. The transition to city life is difficult and sometimes frightening, and there is incontrovertible evidence that many migrants in communally divided societies find a sense of security in wider networks established within their ethnic community. In the case of the Copperbelt in southern Africa, this ethnic identification did not lead to violence. But when other factors – such as wider political tensions or difficulties in finding work and adequate housing – are brought in to play then a vicious and nasty form of warfare can be a result.

Honour and masculinity

The honour of men in Lebanese cities was increasingly determined by the proper behaviour of their women, and by their ability to protect their wives, daughters and sisters from becoming shamed. With the collapse of the extended family, this became an increasingly onerous burden, and the need to control women could create a rebellious reaction that required more controls in a depressing cycle of greater repression.

The ethnic nationalism articulated by such confessional organizations as the Phalanges and Amal militias laid great stress on family values, on women as mothers of the nation, and men as the defenders of their family and community. Discipline, authoritarianism and strong leadership were highly valued, and reflected and justified the patriarch's role in the sexually repressive family. Women were valued as mothers and the providers of hearth and home. And the party and militia structures of ethnic nationalism provided opportunities for young men to behave in an obstreperous and violent fashion not sanctioned by the repressive family, and which, by posing a threat to the women of the Other, enhanced the need for militiamen to protect their own women from assault.

Thus confessional or ethnic organization created a kind of order out of chaos and provided a discourse that explained a difficult and unjust world. It defined a hated Other and gave men someone to blame for the misery of urban life. It provided a sense of community and emotional security in a highly competitive urban economy and society. It supported and justified the repression of women and adolescents by the patriarchal casualties of modernity. And it created a role for young men as the protectors of women in war.

References

Abdennur, A. (1980) 'Combat reactions among a sample of fighters in the Lebanese civil war', *The Psychiatric Journal of the University of Ottawa* 5(2): 125–8.

Accad, E. (1991) 'Sexuality and sexual politics: conflicts and contradictions for contemporary women in the Middle East', in Talpade Mohanty, C., Russo, A. and Torres, L. (eds) *Third World Women and the Politics of Feminism*, Bloomington and Indianapolis: Indiana University Press.

Ajami, F. (1986) *The Vanished Imam: Musa al Sadr and the Shia of Lebanon*, London: I.B. Tauris.

Anderson, B. (1983) *Imagined Communities: Reflections on the Origins and Spread of Nationalism*, London: Verso.

Brown, D. (1994) *The State and Ethnic Politics in Southeast Asia*, London and New York: Routledge.

Campbell, C. (1992) 'Learning to kill? Masculinity, the family and violence in Natal', *Journal of Southern African Studies* 18(3): 614–28.

Chhachhi, A. (1989) 'The state, religious fundamentalism and women: trends in South Asia', *Economic and Political Weekly* 24(18 March): 567–78.

Chhachhi, A. (1991) 'Forced identities: the state, communalism, fundamentalism and women in India', in Kandiyoti, D. (ed.) *Women, Islam and the State*, London: Macmillan.

Das, V. (ed.) (1992) *Mirrors of Violence: Communities, Riots and Survivors in South Asia*, Delhi: Oxford University Press.

Das, V. (1997) *Critical Events: An Anthropological Perspective on Contemporary India*, Delhi: Oxford India Paperbacks.

Entelis, J.P. (1974) *Pluralism and Party Transformation in Lebanon: al-Kata'ib, 1936–1970*, Leiden: Brill.

Epstein, A.L. (1958) *Politics in an Urban African Community*, Manchester: Manchester University Press.

Eriksen, T.H. (1993) *Ethnicity and Nationalism: Anthropological Perspectives*, London: Pluto Press.

Fernea, E. (1985) *Women and the Family in the Middle East: New Voices of Change*, Austin: University of Texas Press.

Fox, R.G. (1997) 'Communalism and modernity', in Ludden, D. (ed.) *Making India Hindu: Religion, Community, and the Politics of Democracy in India*, Delhi: Oxford India Paperbacks.

Fromm, E. (1955) *The Sane Society*, New York: Fawcett.

Gellner, E. (1983) *Nations and Nationalism*, Oxford: Blackwell.

Ghoussoub, M. (1987) 'Feminism – or the eternal masculine – in the Arab world', *New Left Review* 161: 3–13.

Gutman, R. (1993) *A Witness to Genocide: The First Inside Account of the Horrors of 'Ethnic Cleansing' in Bosnia*, Shaftesbury, Dorset: Element Books.

Hanf, T. (1993) *Coexistence in Wartime Lebanon: Decline of a State and Rise of a Nation*, London: Centre for Lebanese Studies and I.B. Tauris.

Hibri, A. al- (ed.) (1982) *Women and Islam*, Oxford: Pergamon Press.

Hobsbawm, E. (1990) *Nations and Nationalism since the 1780s: Programme, Myth, Reality*, Cambridge: Cambridge University Press.

Johnson, M. (2001) *All Honourable Men: The Social Origins of War in Lebanon*, London and New York: Centre for Lebanese Studies and I.B. Tauris.

Khlat, M. (1989) *Les Mariages Consanguins à Beyrouth: Traditions Matrimoniales et Santé Publique*, Évry Cedex: INED/Presses Universitaires de France.

Khuri, F.I. (1972) 'Sectarian loyalty among rural migrants in two Lebanese suburbs: a stage between family and national allegiance', in Antoun, R. and Harik, I. (eds) *Rural Politics and Social Change in the Middle East*, Bloomington: Indiana University Press.

Khuri, F.I. (1975) *From Village to Suburb: Order and Change in Greater Beirut*, Chicago and London: Chicago University Press.

Ludden, D. (ed.) (1997) *Making India Hindu: Religion, Community, and the Politics of Democracy in India*, Delhi: Oxford India Paperbacks.

Makiya, K. (1993) *Cruelty and Silence: War, Tyranny, Uprising, and the Arab World*, London: Jonathan Cape.

Malarkey, J.M. (1988) 'Notes on the psychology of war in Lebanon', in Barakat, H. (ed.) *Toward a Viable Lebanon*, London and Sydney: Croom Helm.

Menon, R. and Bhasin, K. (1998) *Borders and Boundaries: Women in India's Partition*, New Delhi: Kali for Women.

Mitchell, J.C. (1956) *The Kalela Dance*, Manchester: Manchester University Press (Rhodes-Livingstone Papers, No. 27).

Nasr, S. (1985) 'Roots of the Shi'i movement', *MERIP Reports* 15(5): 10–16.

Saad-Ghorayeb, A. (2001) *Hizbullah: Politics and Religion*, London: Pluto Press.

Salibi, K.S. (1976) *Crossroads to Civil War: Lebanon, 1958–1976*, London: Ithaca Press.

Sayigh, R. (1994) *Too Many Enemies: the Palestinian Experience in Lebanon*, London and New Jersey: Zed Books.

Sharabi, H. (1988) *Neopatriarchy: A Theory of Distorted Change in Arab Society*, New York and Oxford: Oxford University Press.

Tabbara, L.M. (1979) *Survival in Beirut: A Diary of Civil War*, London: Onyx Press.

Vanaik, A. (1997) *Communalism Contested: Religion, Modernity and Secularization*, New Delhi: Vistaar Publications.

Younes, M. (1999) *Ces Morts qui nous Tuent: La Vengeance du Sang dans la Société Libanaise Contemporaine*, Beyrouth: Editions Almassar.

Further reading

Johnson, M. (2001) *All Honourable Men: The Social Origins of War in Lebanon*, London and New York: Centre for Lebanese Studies and I.B. Tauris. This book contains an extended treatment of the subject discussed in this chapter.

Sharabi, H. (1988) *Neopatriarchy: A Theory of Distorted Change in Arab Society*, New York and Oxford: Oxford University Press. Hisham Sharabi discusses patriarchal forms of social and political organization in the Middle East as a whole.

Chhachhi, A. (1989) 'The state, religious fundamentalism and women: trends in South Asia', *Economic and Political Weekly*, 24 (18 March): 567–78. Chhachhi, A. (1991) 'Forced identities: the state, communalism, fundamentalism and women in India', in Kandiyoti, D. (ed.) *Women, Islam and the State*, London: Macmillan. Amrita Chhachhi's articles give excellent accounts of patriarchal repression of women in India.

Eriksen, T.H. (1993) *Ethnicity and Nationalism: Anthropological Perspectives*, London: Pluto Press. Thomas Eriksen provides a very useful overview of a number of related issues in a variety of contexts.

10

MOBILIZING THE RHETORIC OF DEFENCE

Exploring working-class masculinities in
the divided city

Karen Lysaght

Summary

This chapter utilizes a gendered lens to examine the issue of youth rioting in Belfast, Northern Ireland, and argues that such activity can be viewed as a form of 'protest masculinity', where young male rioters draw upon a wider hegemonic ideal of their group, that of defensive masculinity. While they attempt to garner resources for themselves within the group through their actions, their ultimate marginality as unemployed, unskilled, poorly educated working-class young men remains unchanged.

- A gendered lens: hegemonic and protest masculinities
- Historical development of a hegemonic ideal of defensive masculinity
- Belfast: a divided city
- Ethnographic snapshots: rioting in interface districts
- Paramilitary orchestration
- Representing rioting: rioters
- Representing rioting: local residents.

Introduction

In a similar situation of long-term ethnic conflict, a gendered perspective illuminates a central cultural theme within the state of Israel. Orna Sasson-Levy (2002) argues that a core value of citizenship is a belief in a 'masculinity of sacrifice', with the pinnacle of this achievement represented by the male combat soldier. While it is not the case that Protestants in Northern Ireland echo this theme, they do have a similar central dynamic which determines many of their rhetorical possibilities and much of the accompanying action. This theme is one which emphasizes a 'masculinity of defence', where all Protestant men must be willing to fight to defend their women-folk, children, communities and ultimately their position within the UK even if their actions should necessitate acting outside the limits of the law.

This represents a hegemonic ideal for Northern Irish Protestant males. Within this chapter, based on research carried out between 1994 and 2002 in working-class communities in Belfast,[1] this central dynamic within the Protestant ethnic group shall be examined through both its historical development and its contemporary articulation, and the ambivalence inherent to these issues shall be illuminated. These contradictory messages shall be explored through an examination of the actions of young men who engage in violence on the boundaries of their communities with similar youths from the Catholic community. A gendered perspective on the actions of these young men shall, therefore, help to illuminate the tensions within this core cultural theme of the Protestant community, in particular drawing upon Bob Connell's (1995) concept of the complex relationship which pertains between hegemonic and protest masculinities.

Rarely has a gendered lens been utilized to examine the masculine nature of the Northern Irish conflict. Those few studies which do focus upon gender usually place their emphasis upon the role of women within embattled communities and their relationship to the instability wrought on families, homes and communities (Sales 1997; Aretaxaga 1997). Given that the conflict has been largely fought by men and the majority of its victims have been male, a more explicit treatment of masculinity is necessitated.[2] In particular, the highly masculinized nature of the public face of the Protestant ethnic group merits attention, where the entire plethora of political, cultural and paramilitary organizations display a decidedly masculine ethos and membership.[3]

Studies of young men in working-class Protestant districts in Northern Ireland (Jenkins 1983; Bell 1990; Gillespie et al. 1992) emphasize their engagement in a range of activities which involve the assertion of their ethnic identity, whether in flute bands, in paramilitary groups or in rioting on the boundaries of residential districts with similar youths from the Catholic ethnic group. Buckley and Kenney (1995) examined incidences of rioting in both Belfast and the large town of Portadown. They found that the violence was recreational in nature, and rarely in their opinion was territory won or lost. Instead the violent exchanges dramatized the relationships of both Protestant and Catholic groups in a carnivalesque fashion to one another, to the apparatus of authority and to their wider social position (Buckley and Kenney 1995). The emphasis of their research, however, focused exclusively upon riots directed by young people against the security forces and did not examine inter-community rioting.

A gendered lens: hegemonic and protest masculinities

The young men who engage in defensive roles within these segregated districts display a social profile which is not dissimilar to that described by Connell when he outlines a marginalized working-class masculinity which he terms 'protest masculin-ity'[4] (1995: 109). Connell views masculinity in relational terms, whereby relations to women but also between groups of men necessitate examination. He outlines the existence of a repressive ideal type of masculinity, which is rarely achieved by the majority of men, but to which men are related in complicit, subordinate and mar-ginalized ways. He labels this ideal type as 'hegemonic masculinity', though one which necessarily varies with social location and circumstance.

This hegemonic form he describes as the complex of historically shaped social practices which attempt to bolster the continuation of the patriarchal oppression of women. The arguments in favour of a patriarchal position usually take the form of arguing that men are superior to women in their intelligence, reason, strength, and ability to provide for their families and defend the vulnerable. The ability of men to actually fulfil all of these varied roles, however, is rarely possible. In fact, Connell notes that few men can actually attain the hegemonic ideal, as they lack sufficient real social power in their lives.

The clearest example of this failure to attain the hegemonic ideal is to be found within Connell's category of marginalized masculinities. Beneath this single label are held some highly disparate groups, which would appear on the surface to share little in common. The first group are gay masculinities, who are clearly marginalized from the hegemonic ideal with its emphasis upon compulsory heterosexuality. The second group are termed 'protest masculinities' and could be easily mistaken for the hege-monic form, given the emphasis upon compulsory heterosexuality and the near casual use of violence. The apparent closeness disguises a fundamental difference, however, which is that these individuals lack any real social power. Instead, they 'pick up themes of hegemonic masculinity in the society at large but rework them in a context of poverty' (Connell 1995: 114).

He notes that for these young men, this represents 'a response to powerlessness, a claim to the gendered position of power, a pressured exaggeration of masculine conventions thereby making a claim to power where there are no real resources for power' (Connell 1995: 111). In fact, their cultural and economic marginality, in terms of unemployment or under-employment and low levels of educational or skill attainment ensure that their claims to any power from the hegemonic patriarchal dividend are 'constantly negated' (ibid.: 116). Instead, they engage in spectacular displays, whereby they actively embrace the marginality and stigma of their position, emphasizing the importance of managing to ensure that they don't lose face, that they keep up a front, in other words that they don't lose respect within their peer group. Haywood and Mac an Ghaill describe this category of protest as the 'exagger-ated observance of a male role' (2003: 39) and the displays emphasized are those which centre on themes of sexuality and violence in what can be described as hyper-masculine performances. There is a common connection of protest masculinities to membership of gangs or other similar groups, which leads Connell to emphasize the importance of the collective aspect of masculinity. While individual practice is necessarily crucial and acknowledged, the group is in fact the bearer of masculinity.

These young men, therefore, make a claim to the dividend of hegemonic masculinity, but largely fail to be rewarded for their attempts. Yet, they also fail to provide any form of resistance to the actual hegemonic constructions which act to marginalize them, as ultimately these young men hold significant continuities with the masculinities to be found in organizations such as the police. Whether they are engaged in resistance behaviour in classrooms (Mac An Ghaill 1994), in the killing of gay men in Australia (Tompsen 2002), in Lebanese gangs in Sydney (Poynting and Noble 1998) or in Bangladeshi gangs in London (Harris 1997), these marginalized young men engage in hypermasculine displays of violence similar to those of the young 'community defenders' of Belfast. The result is that these young men draw upon the hegemonic ideal, yet receive few of the resources associated with this category. In fact, they do not merely maintain their marginalized working-class position, but become further distanced from their own communities through their actions of 'protest'. For Connell, while protest masculinity is clearly an active response to the circumstances these young men find themselves within, and it builds itself upon the ethic of working-class solidarity, it actually results in undermining the connection of these young men to the wider working-class group (Connell 1995: 117), and leaves them ultimately isolated.

Historical development of a hegemonic ideal of defensive masculinity

The history of the development of a hegemonic ideal of the masculine defence of community is not merely a contemporary theme in the north of Ireland, but instead has a long historical record which can be traced back to the origins of the settlement of the most northern parts of the island of Ireland with Protestant settlers from Scotland, Wales and England in the early seventeenth century. Relationships with the displaced Catholic population were characterized by hostility and distrust. As a result, the Protestant settlers lived with a constant disquiet about the possibility of uprisings and attacks on their settlements. Over the centuries which followed, a series of uprisings or rumoured insurrections met with the formation of a series of legal and extra-legal volunteer armies of Protestant men ready to defend their settlements and way of life. The early twentieth century witnessed a recurrence of this in response to a powerful movement toward Irish self-government on the part of Irish Catholic Nationalists. In 1912 approximately 220,000 Protestant men signed the Solemn League and Covenant, pledging to:

> stand by one another in defending for ourselves and our children our cherished position of equal citizenship in the United Kingdom and in using all means which may be found necessary to defeat the present conspiracy to set up a Home Rule Parliament in Ireland.

> (quoted in Bardon 1992: 437)

From the names of those who signed the Ulster Covenant, a volunteer army was recruited, numbering 90,000 men by the end of 1913 and given the name of the Ulster Volunteer Force (Bardon 1992). This group became an armed militia and presented a threatening demeanour in the face of any attempt to grant Ireland self-government, which was viewed as resulting in leaving Protestants as a permanent minority on a largely Catholic island. Instead, the country was partitioned creating a Catholic majority southern state and a corresponding northern state with a Protestant majority. Both entities set up their own statelets and proceeded with the business of self-government.

This division of the island remained a source of unease for Irish Nationalists, and in particular for those who became the permanent Catholic minority within the state of Northern Ireland. They were viewed with suspicion as being politically disloyal and potentially rebellious. As a result, they were granted few social, cultural, economic or political protections, and instead became a highly marginalized population. Over the course of the twentieth century, several episodes of Republican violence erupted, with the most serious and long-term eruption of violent conflict known as 'the Troubles', which spanned from 1969 until 1994. One of the features of the early 'Troubles' was the eruption of inter-communal tension, with attacks on residential property and businesses and the massive dislocation of population across Northern Ireland, as families fled from mixed areas into religiously homogeneous communities (Darby 1971, 1986; Boal 1982). The result was that Northern Ireland became highly segregated, especially in large urban centres such as Belfast.

In addition, both Catholic and Protestant districts saw the formation and growth of vigilante groups whose expressed role was the patrolling and defence of these fragile community boundaries. From these vigilante groups, several of the paramilitary organizations were to develop. These extra-legal organizations were committed to the use of violence to achieve their political aims, whether of reunification of the island of Ireland, or of ensuring the maintenance of the union with Britain. In addition, however, all of these organizations maintained as central to their duties the defence of their home communities. Within the Protestant community, the professed aim of contemporary Loyalist (Protestant) paramilitary organizations such as the Ulster Volunteer Force (UVF)[5] and the Ulster Defence Association (UDA) is to utilize extra-legal violent tactics through which to engage Republican (Catholic) groups such as the Irish Republican Army (IRA).

Others who have appeared to flirt with the use of extra-legal violence include individuals such as the highly prominent Protestant politician and church leader, Dr Ian Paisley. Paisley is a Member of the British Parliament at Westminster, a Member of the European Parliament, the leader of the largest Unionist political party in Northern Ireland, the Democratic Unionist Party and head of the Free Presbyterian Church. Paisley utilizes a form of rhetoric which is highly attuned to the masculinity of defence, with an emphasis on saying 'no' to any form of political compromise, saying 'no' to the possibility of negotiation, or power sharing arrangements with Republicans, saying 'no' to any change in the status of the union with the United

Kingdom. Paisley has on several occasions gathered together Protestant men in displays of the potential of armed resistance in the case of an unpopular political settlement. He has not, however, appeared to overstep the boundaries and utilized extra-legal violence. Instead his actions are a demonstration of a willingness in the event of crisis to resort to extra-legal means to maintain the constitutional *status quo*.

In fact, Ian Paisley is vociferous in his condemnation of Loyalist paramilitarism for their use of illegal forms of violence. An uneasy relationship exists, therefore, between the Loyalist paramilitary organizations and this leading political figure. Between their respective positions, there is disagreement on what constitutes the point of crisis where an extra-legal response is necessitated, but the topic of discussion for both is the same as that debated over several centuries, that of the willingness of Protestant men in Northern Ireland to step outside the boundaries of the law in order to defend their community. Whether those debating this subject are the epitome of societal respectability or indeed those engaged in illegal paramilitary groups in working-class urban districts, the willingness of men to break the law for the purpose of defending Protestant group interests is a cultural theme which has a long historical association with the Northern Irish Protestant population. For a group with a strong tendency to remember and commemorate its past, a theme of defensive masculinity proves both an important rallying cry and a powerful means to enforce conformity.

Belfast: a divided city

Belfast is a highly divided city, where many of the working-class communities are as much as 99 per cent Protestant or Catholic in demographic profile. While some of these districts hold populations of possibly one hundred thousand people, other areas have pockets of one and a half thousand people living in a highly segmented political geography. Since the 1950s, there has been an ongoing process of deindustrialization in the heavily industrialized city of Belfast. Closure and downsizing has been the outcome for many of the traditional workplaces associated with heavy engineering, shipbuilding and the linen and tobacco industries. The employment environment changed to one heavily dependent on employment within the public service, and with the onset of the violent conflict, the security forces. The manufacturing sector that survived or was subsequently established relocated to out-of-town industrial estates.

With the redevelopment of slum housing within the old residential districts associated with the traditional industries and the construction of new suburban estates of public housing, the working-class Protestant population of the city's occupational communities relocated in vast numbers to these new locations. This resulted in leaving Protestant areas within the city highly depopulated with a predominantly older age profile, while Catholic districts, in contrast, remained densely populated and considerably younger in age profile.

In addition to these processes, the violent conflict in Northern Ireland resulted in the entrenchment of segregation between Catholics and Protestants. During the early years of the conflict, tens of thousands of people relocated into districts which held a clear majority of their own co-religionists. In these religiously homogeneous districts, families hoped to live without fear of attack or intimidation within heavily fortified residential districts. Temporary barriers erected in times of crisis between neighbouring districts of differing religious affiliation were replaced over time with more permanent structures; first walls constructed from large sheets of corrugated iron and later with permanent brick walls with adjacent areas of shrubbery. These exact places at which the two 'communities' meet have long provided sites of violence, which mostly takes the form of regular displays of street violence by young men in their teenage years entering into combat with their equivalents in the 'other' community. These displays can at times escalate into more serious incidents involving paramilitary organizations in violent confrontation with one another.

Ethnographic snapshots: rioting in interface districts

Inter-community rioting involves a variety of activities, which range from skirmishes between boys, sometimes as young as three or four, at the boundaries of neighbouring communities, usually involving stone throwing or the use of paint bombs or catapults, up to orchestrated incursions into neighbouring districts by paramilitary-affiliated personnel. While stone throwing can occur between very young boys living adjacent to the interface, this is often mere mimicry of older boys who are the regular protagonists in these displays. The more serious and sustained street fighting usually involves teenage young men, between the ages of fourteen and eighteen. Young women are rarely involved in these exchanges, except in the role of admiring girlfriends or as 'lookouts' for the police.

A basic profile of these young men shows that they are all from working-class communities, usually those immediately adjacent to the site of violence though they might travel from more distant parts of the city at times of political tension. Many of the younger age range are attending secondary school, while the older youths have left full-time education. Many have achieved a very poor standard of education, having displayed problems with truancy and exclusions while at school. They lack any skill base for the labour market, and are often unemployed or engaged in training schemes through the Training and Employment Agency. These young men are often those who refuse to engage in organized activities of youth clubs and community centres, known by some youth workers as 'unclubbable'. When they do attend organized activities they are often disruptive and tend toward regular exclusion from clubs, schools and centres. For these young men, their performances of bravado among their peer group could involve throwing a snooker ball through a youth club window or harassing an inexperienced youth worker. Their most important displays of macho bravado are seen, however, on the boundaries of their home districts.

Here, they engage in demonstrations of masculine prowess, showing that they possess 'a really good arm', being strong and accurate in stone throwing. Indeed, the degree of familiarity with one another that these young opponents have is quite surprising, with young men often organizing an evening of rioting with their equivalent number from the neighbouring district through shouted taunts across streets in the city centre, or at the interface earlier in the day. Regular neighbouring protagonists become acquainted with one another's names from hearing them shouted in the mêlée to fellow protagonists, and can call out targeted pieces of abuse or challenges to named individuals. Indeed, protagonists on both sides gain reputations as good fighters or indeed as those whose performances are poorly evaluated. The criteria against which such judgements are made appear to include a negative appraisal of those whose violence is uncontrolled, or merely aimed at innocent pedestrians. Instead, a good performance is judged to involve direct engagement with neighbouring protagonists.

While there is the potential for serious injury and property damage, the violence is relatively controlled. While young rioters occasionally make remarks to the effect that they were meeting 'the other side' in order for 'their hardest to fight our hardest', the reality is that the violence is usually played out at a sufficient distance to allow protagonists to stand aside when stones or ball bearings approach. The seasonal and temporal nature of rioting is crucial, as the majority of street clashes occur in the bright, dry, long evenings of the summer months. Wet, dark winter nights are guaranteed to be quiet, even on a hotly contested interface. As such, the nature, location and timing of riotous behaviour are known to protagonists, to local residents and indeed to the police.

Serious boundary violence is quite often the result of a spiralling of violence at the local boundary up through various local age cohorts. Should several eight year olds throw stones at a similarly aged group and draw a response, the violence can escalate quickly. From the initial spark, older boys and teenagers could become involved to engage in some of the excitement of the exchange. This can spiral up through various age grades until it is men who become involved in the street violence. The men involved in such roles are usually those who are directly or indirectly involved with the paramilitary organizations operating locally.

Paramilitary orchestration

While the street violence appears to be a predominantly recreational activity providing excitement to those with little financial ability to pursue other recreational possibilities, other more political influences are brought to bear on young rioters. In fact, on occasion, local paramilitary commanders request that the violence should be alternatively escalated or de-escalated depending on the desired result. At times when the paramilitary groups should wish to create some inter-community tension, the young men are instructed to engage in street violence. Alternatively, when it is politically advantageous for it to remain quiet on the interfaces, paramilitary commanders request these young men to desist from any action.

This is referred to by local people as the paramilitaries 'turning on and off the violence' to suit their agenda, whether this be destabilizing the wider political atmosphere or protesting about a particular issue, or indeed justifying the need for funding for a community project to work with young rioters. In all cases, the requests are made by drawing upon the role of the young men as community defenders and as engaged in activities which are of benefit to the political interests of the wider Protestant group. As such, the paramilitary commanders draw upon the hegemonic ideal of defensive masculinity, and locate the actions of the young men clearly within this frame. Their rhetorical stance draws a parallel between the role of the young male rioter and that of the paramilitary organization, as being merely a difference in scale of operation, but ultimately engaged in pursuing the one goal, community defence. Indeed, many of those men within paramilitary organizations in these local areas point out that their initial political activity involved such acts of street violence in defence of the local boundaries.

Representing rioting: rioters

One of the ways in which young people describe their engagement in rioting points explicitly toward the excitement and thrill of the violent exchange. The following extended extract from an interview transcript outlines Jason's experiences and assessment:

> When I grew up that [interface] was the focal point. That's where you went for a fight, that's where you went to throw stones and that's where you would take a risk and maybe go down to the third street in Laurelvale[6] [a Catholic district] when you were sixteen and then run back up again . . . Oh, we'd go down and back up. It was really funny. I used to run around, when I was nine or ten, with an older crowd and I used to be like a wee scout. They would send me down to scout, shout something, and then run like hell just to get the crowd up. I was the youngest and I was more willing to do it at that stage. I didn't really have a clue what was happening. I didn't understand religion or anything. I don't know how many times I had my head split open or got beat up . . . When we were younger we took far more risks. We maybe went into Craig Street and done a bit of graffiti or something. Jesus, you were dead [if they caught you], there's no doubt about it. You weren't getting away with it . . . you were dead. It was the risk and the rush you got after, it was like woah!
>
> (Jason, Protestant male in his mid-twenties)

For young men like Jason, district boundaries provide a source of excitement and thrill-seeking in an environment where positive life prospects are minimal. The activities he describes involve risk-taking, baiting and taunting the other group, openly challenging them to engage in a contest of skill. While this clearly demonstrates the recreational benefits of rioting for highly marginalized young men, it is largely apolitical in orientation. In fact, while young men do emphasize the excitement of rioting, they also locate the same actions clearly within the rhetoric of defensive masculinity.

In discussing the location or nature of rioting they return continuously to their role as community defenders, and depict their actions as the first line of defence for vulnerable interface districts. Without their activities, they contend, the 'other' community would push forward territorially. The young people see themselves as the visible presence of their group on the boundaries of territory continually subject to political contest. Their actions are the demonstration of the continued presence of their group in the area and of the continued maintenance of the boundary by local personnel. As such, they view their role as not divorced from that wider defence of community performed by the paramilitary organizations locally, and indeed local assessments ultimately bolster their assessment.

Representing rioting: local residents

When local inhabitants in their forties and fifties speak of rioting they usually portray it as the actions of bored teenagers, who lack alternative sources of amusement. The violence is mostly viewed as a nuisance, and as merely attracting violence to particular sites. The result of interface violence has often been the depopulation of the immediate area with often imperceptible changes to the location of the interface. Given the overcrowding of Catholic residential areas in Belfast and the underpopulation of Protestant districts,[7] the choice to relocate further into the sanctuary community is not available to many Catholics, thereby keeping a resident population directly on the interface. The result is that interfaces are invariably edging forward into Protestant districts. This, in turn, fuels the sense that territory is being lost and gained through street skirmishes, which is clearly articulated in the following comments by residents in interface areas:

> We've been asking for gates and walls down there for the last five years, and we've been totally ignored . . . at one stage we had eighteen houses empty but now there's about thirty-seven in that area. You see, the way it was, it was methodical. First of all in John's Square itself [small estate of public housing], it's relatively new, it's about ten or twelve years old, and every house on Haight corner [the interface] was occupied. They continually bombarded the first two houses. They ended up being the first two empty. The other house on the far side of the street started to get it, and then they moved up that way and then it started coming up into John's Square, and methodically they came down and put windows in, and windows in, and windows in, until the people couldn't take it no more because they had young families . . . and the intention is that the [Catholic community] want to take over that area and be re-housed and that takes

you up to Marshall Street [a main thoroughfare and marching route for the fraternal organizations] and that creates all sorts of problems.

(Timothy, Protestant male in his late forties)

When's the last time you saw a street that was Catholic and is now Protestant, you just don't . . . you can actually see the interface moving in front of your eyes . . . when the young ones from the other side are standing drinking carry outs [alcohol bought at an off-licence] half a street nearer to you each year, it doesn't seem much, but over a few years . . .

(Alice, Protestant woman, early forties)

In both cases, there is a sense that territory is being lost by the Protestant community through youthful skirmishes, and Timothy makes it clear that he views this process as orchestrated and strategic on the part of the Catholic community. Such sentiments are common in these depopulated Protestant districts, where the actions of young men are viewed as engaged in the active winning and losing of territory. In fact, the language which is utilized to describe street violence as being either the work of bored youths or alternatively motivated by political considerations often takes a sectarian complexion. As such, individuals view young men from their own community as merely suffering from boredom, while those engaged in identical acts from the Catholic community are viewed as motivated by sectarian hatred and orchestrated by political elements. Local residents complain that they feel under pressure to leave their homes, and that the security forces provide them with little protection. As a result, they note that the only factor stopping the progression of Catholic elements into their residential district is the constant presence of the young rioters on the contested boundaries, and by extension dependence on the paramilitary organizations to maintain the dividing line, given any escalation in interface tension. As such, local residents present a highly ambivalent reaction to rioting, which in turn bolsters the assessment of the young men engaged in street violence that they represent the first line of community defence, and thereby articulate through their actions the central hegemonic ideal of defensive masculinity.

Conclusion

The reading of the social and spatial text of riotous behaviour in urban districts of Northern Ireland illuminates the fact that these spectacular displays of extra-legal violence are never merely recreational encounters. While they are indeed displays of group identity (Buckley and Kenney 1995), each act simultaneously tests and asserts territorial claim and counter claim. Every action is met 'with an equal and opposite reaction', whereby continued control of territory is asserted. If the response is weak, land is lost, and with it associated rights to housing, marching, and standing on street corners are conceded. The choice of whether to condemn, condone or indeed encourage defensive acts shifts with context, thereby making visible the necessary tension within the ideological frame of defensive masculinity. Those young men who engage in stone throwing find their actions treated at various junctures as alternatively the work of bored 'rough' youths, the pawns of paramilitary elements who orchestrate acts of politically timed violence through their persons or the first line of defence of the local embattled community.

While the young men in question acknowledge the thrill factor of their actions, they also repeatedly draw upon the rhetoric of defensive masculinity in their own self-presentations. As such, young stone throwers justify their actions and represent their identities within the same terms that Ian Paisley utilizes in his political strategy, and the Solemn League and Covenant articulated in 1912, thereby mobilizing this central hegemonic theme of Northern Irish Protestant group identity. Through their spectacular displays, these young men attempt to draw upon the resources of this central cultural theme, yet their actions meet with local ambivalence and often condemnation, and any rewards gained are merely local and temporary, more often the approval of their peer group, or the admiration of local young women. Like R.W. Connell's 'protest' masculinities (1995), their situation is neither transformed, nor is the hegemonic centre which marginalizes them resisted. Instead, they remain tied to central hegemonic ideals, which emphasize that the source of their victimhood and that of their community lies with an outside aggressor, and that the appropriate response is one of ethnic solidarity in the face of challenge.

Notes

1 The research on which this paper is based was carried out first as part of doctoral research on Loyalism in Belfast and later as part of the British Economic and Social Research Council's Violence Research Programme (VRP) study 'Mapping the Spaces of Fear', which examined the impact of sectarian violence on everyday life in the city.

2 The few studies which do focus explicitly upon masculinity emphasize specific issues such as football (Bairner 1997) or spatial practice (Dowler 2001; Lysaght 2002) but do not consider the masculine nature of the conflict more generally. Several studies have taken a specific interest in young men (Bell 1990; Buckley and Kenney 1995; Gillespie et al. 1992; Jenkins 1983), but they do not emphasize a gendered perspective on their social practices.

3 The fraternal organizations such as the Orange Order and the Apprentice Boys are male only groups, though some have separate women's sections, the flute bands predominantly deny membership to women, and politics until the 1994 cessation of violent activities was largely a male preserve. Paramilitary organizations are made up of 'teams of men', and rarely include female members, though some women might undertake to hide weapons, wash clothes or provide safe houses.

4 Tompsen (2002) provides a list of social characteristics for protest masculinities which include a common pattern of childhood poverty, school failure, petty delinquency, substance abuse, and ongoing marginality in the job market and in housing.

5 The Ulster Volunteer Force is named after the 1912 organization which resisted the movement for Irish Home Rule.

6 All place names have been changed to protect the anonymity of respondents and fieldsite.

7 Over the course of the years of conflict, Protestant districts in Belfast became heavily

depopulated with out-migration to suburban housing estates. The majority of the suburban housing circling Belfast is Protestant in religious make-up and Catholics do not tend to migrate to these new districts, choosing instead to remain in the city, representing a significantly larger proportion of the population in the face of this Protestant 'flight'.

References

Aretaxaga, B. (1997) *Shattering Silence: Women, Nationalism and Political Subjectivity in Northern Ireland*, Princeton, NJ: Princeton University Press.

Bairner, A. (1997) ' "Up to their knees"? Football, sectarianism, masculinity and Protestant working-class identity', in Shirlow, P. and McGovern, M. (eds) *Who are 'The People'? Unionism, Protestantism and Loyalism in Northern Ireland*, London: Pluto Press.

Bardon, J. (1992) *A History of Ulster*, Belfast: Blackstaff Press.

Bell, D. (1990) *Acts of Union: Youth Culture and Sectarianism in Northern Ireland*, London: Macmillan.

Boal, F. (1982) 'Segregating and mixing: space and residence in Belfast', in Boal, F. and Neville, D. (eds) *Integration and Division: Geographical Perspectives on the Northern Ireland Problem*, London: Academic Press.

Buckley, A. and Kenney, M. (1995) *Negotiating Identity: Rhetoric, Metaphor and Social Drama in Northern Ireland Politics*, Belfast: Cadogan Group.

Connell, R.W. (1995) *Masculinities*, Cambridge: Polity Press.

Darby, J. (1971) *FLIGHT: A Report on Population Movement in Belfast During August 1971*, Belfast: Community Relations Commission.

Darby, J. (1986) *Intimidation and the Control of Conflict in Northern Ireland*, Dublin: Gill and Macmillan.

Dowler, L. (2001) 'No man's land: gender and the geopolitics of mobility in West Belfast, Northern Ireland', *Geopolitics* 6: 158–76.

Gillespie, N., Lovett, T. and Garner, W. (1992) *Youth Work and Working-Class Youth Culture: Rules and Resistance in West Belfast*, Buckingham: Open University Press.

Harris, R. (1997) Interpreting teenage inter-ethnic violence, *International Journal on Minority and Group Rights* 4: 301–21.

Haywood, C. and Mac an Ghaill, M. (2003) *Men and Masculinities: Theory, Research and Social Practice*, Buckingham: Open University Press.

Jenkins, R. (1983) *Lads, Citizens and Ordinary Kids: Working-Class Youth Lifestyles in Belfast*, London: Routledge and Kegan Paul.

Lysaght, K. (2002) 'Dangerous friends and deadly foes: performances of masculinity in the divided city', *Irish Geography* 35(1): 51–62.

Mac An Ghaill, M. (1994) *The Making of Men: Masculinities, Sexualities and Schooling*, Buckingham: Open University Press.

Poynting, S. and Noble, G. (1998) ' "If anyone called me a wog, they wouldn't be speaking to me alone": protest masculinity and Lebanese youth in Western Sydney', *Journal of Interdisciplinary Gender Studies* 3(2): 76–94.

Sales, R. (1997) 'Gender and Protestantism', in Shirlow, P. and McGovern, M. (eds) *Who are 'The People'? Unionism, Protestantism and Loyalism in Northern Ireland*, London: Pluto Press.

Sasson-Levy, O. (2002) 'Constructing identitites at the margins: masculinities and citizenship in the Israeli army', *The Sociological Quarterly* 43(3): 357–83.

Tompsen, S. (2002) *Hatred, Murder and Male Honour: Anti-Homosexual Killings in New South Wales, 1980–2000*, Canberra: Institute of Criminology, Research and Policy Studies.

Further reading

Poynting, S. and Noble, G. (1998) ' "If anyone called me a wog, they wouldn't be speaking to me alone": protest masculinity and Lebanese youth in Western Sydney', *Journal of Interdisciplinary Gender Studies* 3(2): 76–94. An examination of inter-ethnic relations in an urban setting utilizing R.W. Connell's concept of 'protest masculinity' through which to understand the gang dynamics of young Lebanese men in suburban Sydney.

Sasson-Levy, O. (2002) 'Constructing identities at the margins: masculinities and citizenship in the Israeli army', *The Sociological Quarterly* 43(3): 357–83. An excellent examination of the interplay of class and ethnicity within the Israeli army, utilizing the concept of 'protest masculinity' in order to understand the actions of working-class, ethnically marginal young Jewish men.

Tompsen, S. (2002) *Hatred, Murder and Male Honour: Anti-Homosexual Killings in New South Wales, 1980–2000*, Canberra: Institute of Criminology, Research and Policy Studies. A detailed examination of the profile of young men engaged in hate crimes in New South Wales. Tompsen provides an illuminating insight into these 'protest masculinities' engaged in regular forays of 'gay bashing' and indeed killing.

11

DEVIANT MASCULINITIES

Representations of neo-fascist youth in eastern Germany

Kathrin Hörschelmann

Summary

In this chapter, I analyse media constructions of neo-Nazi youths in eastern Germany and argue that hegemonic definitions of masculinity rely on a contrast with hyper-masculinity as much as on a rejection of femininity. I point towards the international as well as inter-generational dimensions of neo-fascism and show that neo-Nazis make claims to masculinity that need to be recognized and deconstructed, if racist and neo-fascist tendencies are to be contested.

- Media discourses: young – east German – neo-Nazi – male
- The scale of right-wing extremism in eastern Germany
- A problem of *the other*?

Introduction

The issue with which I am dealing in this chapter is a thorny one. I aim to show how the social legitimacy of western hegemonic masculinities relies on a contrast with an Other who is not so much seen as lacking masculinity, but as possessing and exercising it to excess. Based on an analysis of British media representations of eastern Germany from 1997 to 2001, I argue that a process of spatial distancing and division allows authors to critique 'excessive', violent neo-fascist masculinities, whilst simultaneously confirming the legitimacy and superiority of 'normal', western masculine behaviours. Yet, in deconstructing these representations, I move on a slippery road: the danger is that, in critiquing representations of eastern Germany as dominated by neo-Naziism I unwittingly construct an apology for some of the most abhorrent acts of racial violence committed in an EU country in recent years. This is certainly not the intention. Rather, I argue in this chapter that an analysis of these representations is necessary not only to contest the marginalization and subordination of people in another country, but also to challenge racism and neo-fascism by exposing its links to hegemonic notions of masculinity and by highlighting its international dimension.

Masculinity is based not only on a contrast with femininity, but also on delimiting the borders of what is seen to be 'normal' and what constitutes 'deviant' masculinity (Connell 1995; Whitehead and Barrett 2001). While this deviance is often described in terms of a closeness to femininity, limits are also set on those attributes that are most centrally associated with masculinity. This is mainly due to contradictions in the definition of masculinity itself. The demand on men to be rational and unemotional, for instance, may conflict directly with actions of daring and aggression. Yet both are seen as signifiers of masculinity (Cornwall and Lindisfarne 1994). A spatial practice accompanies the process of defining legitimate masculinities, as described by Phillips in his discussion of the geographies of colonial adventure:

> Unknown, distant spaces of adventure are vehicles for reflecting upon and (re)defining domestic, 'civilized' places. With elements of a recognisable world, recombined in a different order and located somewhere off the edge of the map, on or around the margins of the known world, the geography of adventure corresponds to what drama critic Victor Turner (1969, 1982) calls a liminal space. It is a marginal, ambiguous region in which elements of normal life are inverted and contradictions displayed.
>
> (1997: 13)

In this chapter I argue that eastern Germany, like much of the formerly (state) socialist world, represents just such a space. It is a 'marginal, ambiguous region' in which the contradictions of liberal democracy and hegemonic western masculinity can be discussed at a safe distance from 'home', externalizing problems that would otherwise be seen as contagious and at the heart of one's own. Western representations of neo-fascism in former east Germany allow engagement with what are seen as 'violent excesses' of youthful masculinities, while placing them at some distance from the authorial self, thus confirming the latter's purity and moral superiority. The effort to maintain spatial distance, however, prevents a thorough analysis of the incipient, far reaching and increasingly global networks of racist and neo-fascist activities. Keeping danger at bay, the strategy of projection and disavowal may at the same time lead to misrecognition of racist violence at 'home' and a failure to resist it on a wider scale. In the following analysis of discursive patterns in British media reports about neo-fascism in eastern Germany I show how Other and Self are defined in terms of a binary opposition between youthful immaturity/maturity, irrationality/rationality, savagery/civility, aggression/moderation, and excess/control. I deconstruct these positions not in order to 'talk away' the problems of neo-fascism, but in order to find more effective, mutual ways of resisting it.

Media discourses: young – east German – neo-Nazi – male

The basis of this first section is a survey of articles on eastern Germany published between 1997 and 2001 in the major British broadsheet newspapers the *Independent*, the *Guardian*, *The Times* and the *Daily Telegraph*. The press reports focused mainly on three issues:

1   legal proceedings against former high-profile members of the Communist Party and the Secret Service;
2   doping scandals of former east German athletes; and
3   neo-Nazi attacks on foreigners.

While all three themes place eastern Germany in a distinctly inferior moral position, it is the third which poses the most complex questions since it deals directly with events *after* unification and creates the most abhorrent picture of criminality and incivility. Eastern Germany emerges from this as a young, masculine and neo-fascist place. These images overlap and connote each other: the common association of youth with immaturity and being 'out of control' links neatly with the depiction of aggressive, neo-Nazi masculinity that appears to be equally in need of 'restraint'. 'Youth' as a problem for society is the dominant image evoked here (Griese 2000). It conveniently allows isolation of the issue away from mainstream adult society, yet also provokes generalized concerns about the 'state of the nation', since young people are most often regarded as an indicator of wider social problems (Griffin 1993; Jenks 1996). At the level of place representation, portraying east Germany in terms of negative youthfulness engenders a sense not only of suspicion but also of lacking development compared to the western, liberal model of democracy. Both these notions, of danger and of incompetence, congeal in the image of youthful, neo-Nazi masculinities. In British media representations, the East becomes both a place where masculinity has crossed the lines of the permissible as well as a feminized landscape of helplessness and fear. This image, however, not only labels the represented. It also constructs a role for the author/observer, since the aggressive and irrational behaviour of the Other appears to demand civilization and control by the Self. Just as in colonial times 'it was the white man's duty to civilize and bring order and fair play to wild and exotic distant places' (Bynon 2002: 33), eastern Germany becomes a 'heart of darkness' (see Karacs, *Independent*, 10 September 2000) whose uncivilized savages are in urgent need of help:

> Hermetically sealed in their communist paradise the people in the East were not exposed to alien ways . . . the closest the natives came to strange-looking people was on the beaches of Hungary and Bulgaria . . . These people need help. They need to be immunized against xenophobia, in the

same way that other nations in western Europe have been in the past decades. The treatment? Give them lots of foreigners to look at, to co-exist with.

(Karacs, *Independent*, 23 August 2000)

Creating fear as well as a sense of moral superiority in the reader, journalists paradoxically adopt the very language of masculine aggression when they discuss anti-Nazi strategies, talking of a 'war on racist crimes', 'determination to crack down on the far right' (John Hooper, *Guardian*, 23 August 2000), 'measures to combat the far right' (Imre Karacs, *Independent*, 17 August 2000), or the need of 'a united front' in the 'battle against fascists' (Toby Helm, *Daily Telegraph*, 23 August 2000). Thus it appears that only the strong hand of regulated masculine authority will be able to stop neo-Nazi violence. Two contradictory gender identifications are at play in this discursive construction. On the one hand, eastern Germany is described as a place 'in fear', 'shocked' or 'under threat', while on the other hand, terms like 'horror', 'terror', 'menace' or 'thuggery' portray it as a threatening place. Two headlines from the *Daily Telegraph* illustrate this:

Neo-Nazis are threat to the east, says Schroeder.

(23 August 2000)

It was an evil so monstrous that it defied belief.

(27 January 2001)

The paradox in fact follows an all too familiar narrative logic that validates claims to masculinity: the hero asserts himself by resisting the monstrous Other, while protecting the helpless victim. The gendered dynamics of this fairytale construction are well explained by Mercer:

The destruction of the monster establishes male protagonists as heroes, whose object and prize is of course the woman. But as the predatory force against which the hero has to compete, the monster itself occupies a 'masculine' position in relation to the female victim.

(1994: 45)

Yet, in several articles, it is not so much the East that is seen as under threat, but rather Germany as a whole and the West. This is particularly noticeable when journalists describe general trends, but quickly move on to identify eastern Germany as the main trouble spot (see Toby Helm, *Daily Telegraph*, 8 February 2001). Reports about atrocities committed in eastern Germany emphasize geographical location, whereas the reader often needs a detailed knowledge of Germany in order to be able to identify whether attacks have occurred in western places. The occurrence of neo-fascist attacks and racist politics in the West poses an ideological problem – how to critique something which one had so firmly located with the Other. Imre Karacs from the *Independent* finds her own answer to this: it is only when *even* the prosperous West experiences neo-fascism that it becomes a real threat:

> Many such incidents, however repugnant, are little more than childish pranks, perpetrated by young, complex-ridden East German lads who have never heard of the Third Reich. But the troubling aspect of the latest anti-Semitic outrages is that they are happening in the most prosperous parts of West Germany. And the phenomenon is no longer confined to the under-class.
>
> (8 October 2000)

The reference to age and class lends her argument a certain naturalness and reinforces the sense of moral outrage. Excess and irrationality fit with the description of the Other, but when they 'spill over' to the Self, they threaten the whole order on which the discourse is built. This is particularly problematic when remembering how reluctantly the Nazi-past was confronted in public debates for decades after the Second World War in western Germany. The projection of neo-Nazi threats to the East alone indicates a fear of engagement with that past and an attempt to locate responsibility for it with the Other.

The scale of right-wing extremism in eastern Germany

Academic research on the issue throws up a much more complex and confusing picture than is apparent from media discourse. On the one hand, rates of neo-fascist activity and of xenophobic attitudes are clearly higher among young people in eastern than in western Germany (Wagner 2000; Hurrelmann *et al.* 2002; Wetzels *et al.* 2001). Thus, more east German youths hold the view that there should be less immigration (56 per cent, compared with 46 per cent in western Germany; Hurrelmann *et al.* 2002). Among the most worrying developments is the strength that some neo-Nazi groups have built up in a number of smaller towns and villages, where even so-called 'nationally-liberated zones' have been established (Wagner 2000). Yet, on the other hand there has been a reduction of xenophobic attitudes between 1992 and 1997 to reach levels near or below those of West German youths in 1992 (German Youth Institute Survey; cited in Kleinert and de Rijke 2000). Sturzbecher and Landua (2001) confirm this for the State of Brandenburg and highlight that the group of young people who reject right-wing extremism totally has grown to reach the same level as in Northrhein-Westphalia in 1996. The authors warn that the danger of neo-fascism must not be underestimated but reject the popular thesis of growing right-wing extremism amongst young people in Brandenburg. In their analysis of the survey 'Youth in Brandenburg', Sturzbecher *et al.* (2001) also caution against overestimations of youth delinquency. Again, they demonstrate a decrease in youth violence and in the acceptance of violence amongst young people since 1993, as well as an increase in the willingness to act against violence, which is, however, not directly related to opposing political extremism or xenophobia. According to Sturzbecher et al., there is a group of two to three per cent of extremely violent, predominantly male youths who pose a significant danger to foreigners as well as to other groups of victims. According to the authors, they cannot be opposed with rational argument or civil courage alone, but need to be tackled through combining punishment with developmental support. Sturzbecher et al. explain that there is no one cause of violence but a whole spectrum of developmental conditions/social contexts, personal characteristics and political opinions that all need to be considered to explain youth violence. Importantly, the authors also demand greater differentiation between the phenomena of violence, right-wing extremism and xenophobic attitudes. This is an important issue highlighted further by Farin and Weidenkaff (1999) and by Wetzels et al. (2001):

> Attitudes towards foreigners, right-wing oriented voting behaviour, membership in right-wing parties, their strategies of recruiting followers, extreme rightwing skinhead and hooligan subcultures, youth violence and crimes that are registered as extremely right-wing are not interchangeable phenomena, even if there are without doubt connections between them.
>
> (Wetzels et al. 2001: 11; my translation)

Media representations of the rise of neo-fascism reflect insufficiently on these distinctions. At the same time, they ignore facts that do not fit into the picture of neo-fascist east Germany or that complicate the simple equation of right-wing attitudes with extremism, thus creating an oversimplified picture which helps little to understand the causes of neo-fascism. Thus, readers will find little mentioning of the fact that the vast majority of east German youths position themselves on the left and in the middle of the political spectrum and identify themselves more strongly with left-leaning politics than west German youths (Hurrelmann et al. 2002). I have also yet to find an article that considers the age dimension of right-wing extremism, which Friedrich (2001) finds to be significant. According to the two surveys consulted by him, anti-foreign sentiments are strongest amongst young people in eastern Germany, while in western Germany, it is the older cohort of respondents (age 40+), who declare that they totally or mostly agree with the statement 'I don't like the many foreigners' (Table 11.1). Concluding that '[i]ntolerance is a phenomenon

Table 11.1 'I don't like the many foreigners' – respondents who either 'totally' or 'mostly' agree (per cent, October 2000)

	14–19	20–9	30–9	40–9	50–9	60+
West	31	28	24	38	41	46
East	44	41	33	36	29	32

Source: Friedrich (2001).

of the young generation in east Germany and of the old generation in west Germany' (20, my translation), Friedrich unsettles the popular claim that east German right-wing extremism is a result of GDR education and socialization.

The gender dimension of neo-fascism is also most frequently presented without further reflection. Although most articles mention that those who commit racist and neo-Nazi crimes are male in their majority, they refrain from looking more closely at the connections between masculinity and violence. Thus, the reader is neither made aware of the way in which aggression and control form part of dominant definitions of masculinity (Cornwall and Lindisfarne 1994; Kimmel 1994), nor of the links between masculinity and gendered conceptions of national identity (Mayer 2000). Yet, claims to hegemonic masculinity on the one hand and assertions of 'protecting the nation' on the other congeal in neo-fascist ideology and can thus motivate extremist actions:

> It is men who are generally expected to defend the 'moral consciousness' and the 'ego' of the nation. Men tend to assume this role because their identity is so often intertwined with that of the nation that it translates into a 'personalized image of the nation' (Hroch 1996: 90–1). Because men 'regard the nation – that is themselves – as a single body' (ibid.), their own 'ego' becomes at stake in national conflicts, and they frequently seek to sustain control over reproduction and representation of both sexuality and nation and over boundaries of the nation, through defining who is included in, or excluded from it.
>
> (Mayer 2000: 6–7)

In neo-fascist attacks, the perpetrators not only enforce a particular, racist ideology of national belonging, but at the same time perform a specific type of masculinity, which 'for most men . . . is central to achieving entry to, and being accepted within, any particular "community" of men' (Whitehead and Barrett 2001: 20). There are parallels here between western assertions of hegemonic masculinity and east German neo-Nazis' claims to masculinity that need to be recognized. Both assert themselves against an Other (or groups of Others), who is deemed inferior, yet 'monstrous' and in need of defeat. Because of this parallel, analyses of right-wing extremism and neo-fascism need to reflect critically on dominant definitions of masculinity and on the methods through which hegemony is constructed.

The flip-side of the coin, however, is where to place women in this equation. In media representations of east Germany, they disappear almost completely. There is little mention of the fact that only 20 per cent of the members of extremist groups are women and that rates of xenophobic attitudes tend to be (at least somewhat) lower for women (Kleinert and de Rijke 2000; Rommelspacher 2000; Hurrelmann et al. 2002). This gender blindness serves to reinforce the masculinized image of eastern Germany. Yet, on the other hand, there is also insufficient information about women's involvement in neo-Nazi groups and about their nonetheless significant support for extreme right-wing policies. The gendered assumption of female innocence and political igno-rance/passivity is confirmed here without considering that rejections of violence amongst women do not automatically spell greater 'tolerance'.

With respect to both male and female east German youth, what is quite clearly missing in the media discourse is a recognition of diversity. The full extent of sub-cultural and political orientations is rarely clarified, nor do we find any consideration of what issues other than nationalism and racism concern young people. Karig and Schuster (1999) begin to give an impression of this in their research from 1995, yet there is surprisingly little work looking in detail at the wider spectrum of youth (sub)cultures in eastern Germany. In the survey conducted by Karig and Schuster, 4.9 per cent of young East German respondents showed 'sympathies' with 'Faschos/Republicans' (neo-Nazis), 6.4 per cent with 'Skinheads' and 24 per cent with 'Heavy Metal Music' (the latter two are not necessarily connected to neo-Nazism), compared with 35.8 per cent for the 'Peace Movement', 58.1 per cent for 'Environmental Protection', 43 per cent for the 'Anti-Nuclear Movement' and 24.1 per cent for the 'Women's Movement'. Punks received 16.4 per cent, the 'Autonomous Left' 12.3 per cent, 'Goths' 11.1 per cent and 'Squatters' 14.1 per cent. Of course, none of the latter categories allow easy conclusions about attitudes towards foreigners and other socially marginalized groups, but they give an impres-sion of the wider political and cultural concerns of young people in eastern Germany. A recognition of this diversity is important particularly in order to avoid simplistic conclusions about the causes of neo-fascism, such as the frequently voiced opinion that a lack of 'civil values' and authoritarian upbringing are the main reasons for the rise of the extreme right in eastern Germany (see Lucian Kim, *Independent*, 13 August 2000 and Imre Karacs, *Independent*, 17 August 2000)

The crucial point here is that, for the supposed lack of civil values and poor education, liberal democracy can be presented as a 'cure' without too much self-reflection. Western, liberal education is deemed to hold the key to greater tolerance and democratic citizenship. Unemployment and the difficult socio-economic situ-ation in eastern Germany are mentioned in some reports, but often discounted on the grounds that boys from middle-class backgrounds are also found in neo-Nazi groups. Socio-economic hardship, however, is an issue that draws into question the successes of liberal market society in providing an adequate standard of living without which the exercise of 'freedom' becomes little more than a hollow phrase (Lash 1995). As Fulbrook argues:

> young people may have freedom, but it is the freedom to wander the street, to stare forelornly into run-down ice-cream cafes or supermarkets, offering a range of goods beyond their pockets. While the vast majority of young people would abhor acts of racism and violence, it is in this soil that, among an ever more active minority, neo-Nazism has been able to put down distinctive roots.

> (1999: 225)

Deprivation combined with male gender, political orientation, lack of parental support, low levels of social integration and low degrees of educational attainment have been identified as key characteristics of young people who join neo-fascist organizations (Sturzbecher *et al.* 2001; Wetzels *et al.* 2001; Griese 2000). Unemployment was at a level of 20.5 per cent for young people between the ages of 20 and 24 in 1998 (higher than the east German average; Statistisches Bundesamt 2000), and places for vocational training are limited. Under these conditions, future career chances look bleak indeed for a large number of young people. Add to this the collapse of youth cultural infrastructure after unification (Starke 1995) and the insecurities caused by socio-political change and it is not too hard to see why certain groups become vulnerable to neo-fascist ideology, which offers them easy answers, targets and a sense of identity. This argument was most clearly expressed in a reader's response to a previous article in the *Guardian*. A 16-year old girl from Halberstadt, near Magdeburg in eastern Germany, wrote:

> I do not deny that neo-fascism and neo-Nazism are problems in Germany. I have had some experience of them and I know only too well that Nazi violence is the worst thing that could happen to anybody.
>
> But the problem is not the young Nazis, but the old people, who experienced the Hitler time and the second world war. They repeat all the old nonsense about the Jews and foreign people. Even my grandparents do and I feel ashamed.
>
> . . . It is easy to blame the young people and not the people who have authority and the power to do something about the Nazi problem.
>
> I, for example, live in the state of Saxony-Anhalt. We have an unemployment rate of about 20 per cent. Boredom and hopelessness force us to find somebody to blame for all the trouble. I do not want to justify the Nazis and the right wing. On the contrary, I condemn everything they do. I blame the German government for not caring about us and not doing anything for us.
>
> (*Guardian*, 23 January 1998)

This letter summarizes several critiques at once, without trivializing the problem of neo-fascism. With reference to the difficult socio-economic situation it asks whether 'young people' are the only ones to blame. The question of responsibility is extended to the older generation and to governmental authorities. It can be taken even further by interrogating the background of neo-Nazi propaganda – who produces it and how does it come to circulate? The German weekly newspaper *Die Woche* published an interesting report in this respect on 17 August 2001. Warning that the main danger emanating from right-wing extremism was its increasing ability to network and to make international connections, the paper identified five intersecting 'layers' of 'Germany's brown Mafia', including illegal organizations, networkers, militants, moderates and the National Democratic Party (NPD). For each of these groupings, the paper identified not only who the main perpetrators of violence were, but also their leaders, ideological backgrounds, organizational structure, media, means of subcultural influence and business links. The connections highlighted here lead far beyond Germany, to international racist organizations including the notorious 'Blood and Honour'. The report also showed up some of the most important sources of material support for neo-fascist networks. While these elements are certainly not sufficient to explain racist violence in eastern Germany, they draw a wider and more complex picture that needs to be understood, if anti-Nazi measures are to succeed. Focusing on those who commit neo-fascist crimes alone will not reveal or destroy the wider networks that sustain them. Significantly, *Die Woche* also called to task governmental politicians, arguing that '[s]o long as Germany's opinion-shaping elite swings arbitrarily between hysteria, trivialization and silence, nobody should be surprised, if the brown fire spreads and is inflamed again and again'.

A problem of *the Other*?

The focus on so-called mainstream politics is an important one, since it is, of course, inaccurate to associate racism and right-wing nationalism only with extremist groups. Both west and east Germany show low levels of voters' support for extreme right-wing parties, as was evident in the parliamentary elections of 2002, when the highest scores reached by an extreme-right party were 1.5 per cent for the NPD and 1.0 per cent for the Republican Party in Saxony on second votes (v. Schwartzenberg 2002). Yet, this may simply be a reflection of the move of mainstream parties to the right to capture voters, as can be observed in many European countries at the moment. By separating 'extremist' from 'mainstream' politics, overlaps between them are hidden, while a certain degree of xenophobia and racism is presented as 'normal' and 'acceptable'. A telling example of this was the case of Edgar Griffin's expellation from the British Conservative Party. His racist views were found to be no longer acceptable under the party line, yet there were those who thought he was simply an 'eccentric with dotty ideas' and generally harmless:

> In my 22 years as a Tory candidate and MP, I came across hundreds of silly old buffers like Edgar Griffin . . . [O]ne has to remember that his views were once the norm and that in every other respect he was an upstanding local citizen. So let us not get too excited by the fact that the party still has some eccentrics with dotty ideas. Most of them are harmless and 'know in their heart that their views are wrong'.
>
> (Michael Brown, quoted in *The Week*, 1 September 2001)

As is clear in this example, different perspectives are applied to similar phenomena, depending on the position and origin of those concerned. We hit here on what is perhaps the most worrying aspect of the media discourse about right-wing extremism: it locates it safely on the side of the Other, while trivializing the rise of xenophobic attitudes and politics in the West. Media reports about the election successes of extreme right-wing parties in France, Austria, Belgium, the Netherlands, Denmark, Italy, Britain and other European countries portray a brief sense of 'concern', followed by reassurances to the reader that these developments are isolated, uncharacteristic and short-term. The French electorate are lauded for standing up to show resistance after Le Pen's success in the first round of the Presidential elections in 2002, while significant reductions in the share of votes for Jörg Haider's Freedom Party in 2002 are interpreted as a sign that the days of right-wing extremism are numbered:

> in the end, Mr Le Pen was soundly defeated. He was, in fact defeated more soundly – 82 per cent to 18 per cent – than any other politician has been defeated in a large democracy in modern times. On 5 May, France stood up for democracy, for tolerance, for Europe and for common-sense.
>
> (Lichfield, *Independent*, 28 December 2002)

Importantly, there is no condemnation of French or Austrian society as a whole, no questioning of their 'civil values' and indeed a distinct effort to play down the remaining strength of extreme right-wing parties. Reports describe the profile of individual leaders like Pim Fortuyn, Jörg Haider and Le Pen at length, thus individualizing the issue rather than examining changes in society that may have led to the rise of xenophobic sentiments and critiquing the role of media and mainstream parties' political discourse (Fekete 2001). Yet, anti-foreign sentiments have been found to be high (with temporal variations) in countries like France and Australia. In the French post-Presidential election survey of 1995 no less than 74 per cent of French respondents agreed completely with the statement 'there are too many immigrants in France' (Lubbers and Scheepers 2002), while Australian Election Studies showed that post-war opposition to immigration in Australia reached a peak in 1993, with 70 per cent of respondents believing that immigration had 'gone too far' (Gibson et al. 2002). As tempting as it may be to locate the rise of extreme right-wing groups and xenophobic attitudes in places far removed from western liberal democracies, these surveys as well as the recent election results in many EU countries show that there is no cause for complacency.

Although the U.K. has not seen major electoral successes of extreme right-wing parties, worrying tendencies can be registered there as well. Not only has the British National Party made significant gains in the local elections of 2002, but racist incidents are frequent, although often unreported in the press. According to data by the Policy Studies Institute nearly 300,000 racist incidents occurred in England and Wales in a 12-month period between 1993 and 1994. Of these, 80 per cent consisted of racist abuse and insults (Virdee 1997; see Chahal 1999). Athwal reports 91 cases of 'people known to have lost their lives to racism, either in racist attacks, in state custody or as a result of seeking asylum, between February 1999 and April

2001' (2001: 117), while Kundnani (2001) and Grandon (2001) report on a series of violent attacks against 'asylum seekers' in Dover, Sheffield, Newcastle, Coventry, Sunderland and Glasgow. According to a Metropolitan police study, there were 18,253 reported racist incidents in London in 2000–01 (*Guardian*, 22 February 2002). Finding detailed information about these incidents in the press, however, is no easy task. Many are not reported nationally (e.g. *London Local*, 30 October 2002, 15 December 2002, 3 January 2003), while others are only mentioned briefly in short news sections (see *BBC News*, 10 October 2002). Yet other cases are reported without any examination of the background of offenders or suggestions of political and social causes (e.g. *BBC News*, 6 January 2003, *Guardian*, 15 December 2001). Brief suggestions are made about institutional racism (*Guardian* 8 December 2000) and failures of the asylum system (*Guardian*, 15 December 2001), but the reader will look in vain for a deeper analysis of socio-political conditions.

Comparing reports about neo-fascist incidents and right-wing extremism in eastern Germany with those about similar occurrences in western Europe shows a divided logic that cannot be easily justified. While neo-fascism amongst young people in east Germany is (rightly) criticized, the xenophobia and racism of 'mainstream' society as well as of radical groups in the West is portrayed as isolated, inconsistent and resisted. Rarely do journalists make the link to wider social and political changes or critically interrogate what might be wrong with their societies. The danger of this dual representational strategy is that xeno-racism can and has become part of an accepted everyday reality in many western democracies today without much public contestation. The strategy of disavowal discussed at the beginning of this chapter thus works in an insidious way not only to delegate all evils to the Other, but also to ignore problematic developments in the Self.

Conclusion

This chapter has dealt with the construction of masculinity in a highly problematic area. Its main aim has been to demonstrate how the formation and maintenance of hegemonic masculinity relies on a contrast not only with women, but also with male/masculinized Others, who are seen as exercising masculinity in excess. In analysing representations of neo-fascism in eastern Germany, my intention has not been to erase from view or trivialize the serious issue of xenophobia and right-wing extremism, but to complicate the assumption of a dualistic divide between East and West that allows externalization and disavowal of those attributes which could disrupt hegemonic masculinity's legitimacy. Against the contrast of a deviant Other, western, middle-class, liberal masculinity appears as rational, responsible and 'in control'. These connotations are destabilized, when attributes of the Other appear at the heart of the Self. Yet, the discussion of neo-fascism in east Germany and in the West also complicates theoretical assumptions, in particular the 'hegemonic masculinity' thesis. In much of the literature, non-hegemonic masculinities are contrasted dualistically with those that are deemed to be hegemonic. Neo-fascist masculinities, however, square this assumption, since they assert their hegemony over others even as they are excluded from dominant definitions of liberal, western masculinity. They do not fit easily into the picture of weak, maginalized Others. Certainly, the deconstruction of hegemonic masculinity cannot mean a reassertion of neo-fascist masculinities, but needs to aim at transforming the former while contesting the latter and at recognizing the dangers that emanate from ignoring the presence of neo-fascist and xenophobic tendencies in the Self. So long as certain forms and expressions of racism are seen as extremist and others as 'normal' and acceptable, fascist ideologies will find fertile soil, especially if 'mainstream' politics appears to confirm many of their claims. At the same time, it is important to recognize diversity in places like eastern Germany in order to understand why some individuals come to associate themselves with neo-fascist groups, while others live their lives across a much wider spectrum of (sub)cultural and political interests. Strengthening the position of the latter and recognizing what it is that prevents them from joining neo-Nazi groups has to be an important element in resisting neo-fascism.

References

Athwal, H. (2001) 'The racism that kills', *Race & Class* 43(2): 111–23.

BBC News (2002) 'Scale of race problem revealed', 10 October.

Bynon, J. (2002) *Masculinities and Culture*, Buckingham: Open University Press.

Chahal, K. (1999) 'The Stephen Lawrence Inquiry Report, racist harassment and racist incidents: changing definitions, clarifying meaning?' *Sociological Research Online* 4(1). Online. Available at http://www.socresonline.org.uk/socresonline/4/lawrence/chahal.html (accessed 29 February 2004).

Connell, R.W. (1995) *Masculinities*, Berkeley: University of California Press.

Cornwall, A. and Lindisfarne, N. (1994) 'Dislocating masculinity: gender, power and anthropology', in Cornwall, A. and Lindisfarne, N. (eds) *Dislocating Masculinity. Comparative Ethnographies*, London and New York: Routledge.

The Daily Telegraph (2001) 'It was an evil so monstrous it defied belief', 27 January.

Die Woche (2001) 'Das Netz der Nazis', 17 August.

Farin, K. and Weidenkaff, I. (1999) *Jugendkulturen in Thüringen*, Bad Tölz: Tilsner Verlag.

Fekete, L. (2001) 'The emergence of xeno-racism', *Race & Class* 43(2): 23–40.

Friedrich, W. (2001) 'Ist der Rechtsextremismus im Osten ein Produkt der autoritären DDR?', *Aus Politik und Zeitgeschichte* B46: 16–23.

Fulbrook, M. (1999) *German National Identity after the Holocaust*, Cambridge: Polity Press.

Gibson, R., McAllister, I. and Swenson, T. (2002) 'The politics of race and immigration in Australia: One Nation voting in the 1998 election', *Ethnic and Racial Studies* 25(5): 823–44.

Grandon, V. (2001) 'Glasgow: a dossier of hate', *Race & Class* 43(2): 123–8.

Griese, H.M. (2000) *Jugend(sub)kultur(en) und Gewalt. Analysen, Materialien, Kritik. Soziologische und pädagigische Beiträge*, Münster, Hamburg and London: Lit-Verlag.

Griffin, C. (1993) *Representations of Youth: The Study of Youth and Adolescence in Britain and America*, Cambridge: Polity Press.

The Guardian (2002) 'Stabbing boy gets five years', 26 October.

Hanke, M.-A. (1998) 'My friends aren't Nazis', *Guardian*, 23 January.

Helm, T. (2000) 'Neo-Nazis in Germany gain 10 pc', *Daily Telegraph*, 23 December.

Helm, T. (2000) 'Neo-Nazis are threat to the east, says Schroeder', *Daily Telegraph*, 23 August.

Helm, T (2001) 'Young Germans see "good side" to Nazis', *Daily Telegraph*, 8 February.

Hooper, J. (2000) 'German prosecutor steps up war on racist crimes', *Guardian*, 23 August.

Hopkins, N. (2002) 'Fifth of racist crime involves neighbours', *Guardian*, 22 February.

Hroch, M. (1996) 'From national movement to the fully-formed nation: the nation-building process in Europe', in Balakrishnan, G. (ed.) *Mapping the Nation*, London: Verso.

Hurrelmann, K., Albert, M. and Schneekloth, U. (ed.) (2002) *Jugend 2002: Zwischen pragmatischem Idealismus und robustem Materialismus*, Shell Jugendstudie, Frankfurt am Main: Fischer Taschenbuch Verlag.

Jenks, C. (1996) *Childhood*, London and New York: Routledge.

Karacs, I. (2000) 'Schröder pledges £25m to combat neo-Nazi attacks', *Independent*, 17 August.

Karacs, I. (2000) 'A cure for Germany's race hate', *Independent*, 23 August.

Karacs, I. (2000) 'Black doctor challenges German racism', *Independent*, 10 September.

Karacs, I. (2000) 'Anti-Semitism "respectable" again as more Jews move into Germany', *Independent*, 8 October.

Karig, U. and Schuster, K.-D. (1999) 'Abseits oder mittendrin? Zu jugendkulturellen Szenen und Stilen', in Bien, W., Kuhnke, R. and Reißig, M. (eds) *Wendebiographien*, Opladen: Leske & Buderich.

Kim, L. (2000) 'Trial of German skinheads who kicked immigrant to death leaves widow in fear', *Independent*, 13 August.

Kimmel, M. (1994) 'Masculinity as homophobia: Fear, shame and silence in the construction of gender identity', in Brod, H. and Kaufman, M. (eds) *Theorizing Masculinities*, Thousand Oaks, London, New Delhi: Sage Publications.

Kleinert, C. and Rijke, J. de (2000) 'Rechtsextreme Orientierungen bei Jugendlichen und jungen Erwachsenen', in Schubarth, W. and Stöss, R. (eds) *Rechtsextremismus in der Bundesrepublik Deutschland: Eine Bilanz*, Bonn: Bundeszentrale für politische Bildung.

Kundnani, A. (2001) 'In a foreign land: the new popular racism', *Race & Class* 43(2): 41–60.

Lash, S. (1995) 'Reflexivity and its doubles: Structure, aesthetics, community', in Beck, U., Giddens, A. and Lash, S. (eds) *Reflexive Modernization. Politics, Tradition and Aesthetics in the Modern Social Order*, Cambridge and Oxford: Polity Press.

Lichfield, J. (2002) 'Far-right tide was turned, but voters remain frustrated', *Independent*, 28 December.

London Local (2002) 'Race attack man fights for life', 30 October.

London Local (2002) 'Man half-blinded in racist attack', 15 December.

London Local (2003) 'Man attacked by racist thug', 3 January.

Lubbers, M. and Scheepers, P. (2002) 'French *Front National* voting: a micro and macro perspective', *Ethnic and Racial Studies* 25(1): 120–49.

Mayer, T. (2000) 'Gender ironies of nationalism: setting the stage', in Mayer, T. (ed.) *Gender Ironies of Nationalism*, London and New York: Routledge.

Mercer, K. (1994) *Welcome to the Jungle*, London and New York: Routledge.

Phillips, R. (1997) *Mapping Men and Empire. A Geography of Adventure*, London and New York: Routledge.

Rommelspacher, B. (2000) 'Das Geschlecherverhältnis im Rechtsextremismus', in

Schubarth, W. and Stöss, R. (eds) *Rechtsextremismus in der Bundesrepublik Deutschland: Eine Bilanz*, Bonn: Bundeszentrale für politische Bildung.

Schwartzenberg, M. v. (2002) 'Endgültiges Ergebnis der Wahl zum 15. Deutschen Bundestag am 22. September 2002', in Statistisches Bundesamt (ed.) *Wirtschaft und Statistik* 10: 823–37.

Seenan, G. (2001) 'Life for murder of Kurd refugee', *Guardian*, 15 December.

Statistisches Bundesamt (2000) *Datenreport 1999: Zahlen und Fakten über die Bundesrepublik Deutschland*, Wiesbaden: Statistisches Bundesamt.

Starke, U. (1995) 'Young people: lifestyles, expectations and value orientations since the Wende', in Kolinsky, E. (ed.) *Between Hope and Fear*, Keele: University Press.

Sturzbecher, D. and Landua, D. (2001) 'Rechtsextremismus und Ausländerfeindlichkeit unter ostdeutschen Jugendlichen', *Aus Politik und Zeitgeschichte* B46: 6–15.

Sturzbecher, D., Landua, D. and Shashla, H. (2001) 'Jugendgewalt unter ostdeutschen Jugendlichen', in Sturzbecher, D. (ed.) *Jugend in Ostdeutschland: Lebenssituation und Delinquenz*, Opladen: Leske und Buderich.

Virdee, S. (1997) 'Racial harassment', in Modood, T., Berthoud, R. and Lakey, J. (eds) *Ethnic Minorities in Britain*, London: Policy Studies Institute.

Wagner, B. (2000) 'Rechtsextremismus und Jugend', in Schubarth, W. and Stöss, R. (eds) *Rechtsextremismus in der Bundesrepublik Deutschland. Eine Bilanz*, Bonn: Bundeszentrale für politische Bildung.

The Week (2000) 'Edgar Griffin: just a typical Tory?', 1 September.

Wetzels, P., Fabian, T. and Danner, S. (2001) *Fremdenfeindliche Einstellungen unter Jugendlichen in Leipzig*, Münster, Hamburg, Berlin and London: Lit-Verlag.

Whitehead, S.M. and Barrett, F.J. (2001) 'The sociology of masculinity', in Whitehead, S.M. and Barrett, F.J. (eds) *The Masculinities Reader*, Cambridge and Oxford: Polity Press.

Further reading

Cornwall, A. and Lindisfarne, N. (eds) (1994) *Dislocating Masculinity: Comparative Ethnographies*. London and New York: Routledge. This edited collection provides an excellent overview of ethnographic work on masculinities. The chapters are both empirically rich and theoretically sound.

Mac an Ghaill, M. (1994) *The Making of Men: Masculinities, Sexualities and Schooling*, Buckingham: Open University Press, remains for me the most compelling and insightful study of the construction of young masculinities.

Whitehead, S.M. and Barrett, F.J. (eds) (2001) *The Masculinities Reader*, Cambridge and Oxford: Polity Press. I would recommend this reader to students looking for a quick but comprehensive guide to core texts on masculinity from across the social sciences.

The journals *Race & Class* and *Ethnic and Racial Studies* offer excellent up-to-date factual information, academic analyses and expert commentaries on racism and neo-fascism in Britain, Europe and other parts of the world.

Part 4

EMBODIED MASCULINITIES

12

SHIFTING SPACES OF MASCULINITY
From Carisbrook to the MCG

C. Michael Hall

Summary

In Australia and New Zealand male urban spaces have historically focused on the pub and the sports field. The pub was for long a male bastion under law, while certain codes and sites of sport were also long excluded from female participation. However, over the past 20 years such spaces have been substantially renegotiated within the context of changing institutional and societal mores and values. The chapter aims to integrate masculine practice with certain theoretical notions of gendered space and its negotiations and does so within the context of the 'ocker' and historical representations and identities of the Australian and New Zealand male.

- Gender and sports spaces
- Sport as a gendered institution
- The micropolitics of gender relations.

Introduction

Access and use of space is inexorably tied to issues of gender, class and power. However, it is also dynamic. In the postcolonial settings of Australia and New Zealand the nature of space and the identities and associations with which it intersects are continuously being renegotiated and reinterpreted. The purpose of the present chapter is to interrogate the shifting notions of gender in relation to an identified 'traditional' space of masculinity – the sportsground.

Sport has long been recognised as significant for gender representation and identity. Sport plays an important role in the gendered and colonial histories of New Zealand and Australia (Phillips 1984; Nauright and Broomhall 1994; Nauright and Chandler 1996; Andrewes 1998; Daley 1999), as well as in accounts of the emergence of regional and national identities (Fougere 1989). However, it must be recognised that post-colonial national identities of the ANZAC, the 'digger', the bronzed Aussie on the Beach at Bondi were essentially male representations of

155

national identity. As well as war and the pioneer experience, such representations found themselves based primarily on sporting prowess and activities and although male oriented, both countries utilised sport as a means of declaring their cultural and political independence from Britain. In the case of Australia this was primarily through cricket and in the case of New Zealand through rugby union. Such attitudes linger to the present day. For example, in early 2003 Australia beat England 3–1 in an international soccer match in England for the first time. Australian supporters at the Upton Park ground where the match was played taunted the English with cries of 'just like the cricket' and 'just like the tennis', in reference to Australia's series defeat of England in the Ashes test series and in a Davis Cup tennis game. In all cases the sports were male.

Nevertheless, to portray antipodean sport as grossly male dominated would, arguably, be an exaggeration. While male sports still tend to dominate sponsorship and media coverage – and attract the biggest crowds – there have been substantial shifts in image, media coverage, and representation in recent years which have challenged notions and spaces of masculinity as well as other contested subjects such as race. Accompanying such changes has been the development of new spatial articulations of gender with respect to both existing space and the development of new space. This chapter discusses such changes in relation to competitive and professional sports space, not just the playing field but also the stadia and facilities that encompass the playing area. In order to illustrate some of the points made in the text with respect to the changing gendered nature of sports space in Australia and New Zealand utilisation is made of the author's field notes of conversations with spectators while attending sporting events in the two countries. Interviewees were primarily male and covered a wide range of age groups. However, the selection of interviewees is perhaps best described as random in that it depended on what seat or section of a stand was allocated through ticket purchase. Names, even if they were offered, were not used while ages, if not provided, were estimated as a result of the development of 'naturalistic' sporting conversations. Conversations were not taped but key quotes or points were written down in the pages of the *Football Record* or other match programmes. Stadia included the Sydney Cricket Ground, the Melbourne Cricket Ground, The Gabba (Brisbane) and Subiaco Oval (Perth) in Australia and Carisbrook (Dunedin), the old Wellington rugby stadium and Lancaster Park (Christchurch) in New Zealand during the period 1998 to 2002.

Gender and sports spaces

Gender is an organising principle in sport, and in the production of the built environment. These are interconnected in the production of sporting and leisure spaces (Bond 1998). Although Australia and New Zealand have long had schooling policies which encourage sport these tended to be segregated, until very recently, by the nature of the sport. For example, netball and, to a lesser extent, field hockey, were female sports while, football (of whatever variety) and cricket were primarily male. Interestingly, tennis and swimming were sports in which both males and females took part, although it was not until the last 20 years that surf lifesaving clubs began to accept female members. Moreover, sporting competitions were, and to a great extent still are, segregated by sex. Inclusion and exclusion in various sports was often divided on the basis of class, e.g. in several Australian states rugby union was long the domain of wealthier private schools. With respect to gender, females were provided for participation in sports which tended to be regarded, by men, as not so physically or intellectually demanding. However, exclusion in participation also meant spatial exclusionary practices. As Massey (1993: 179) recognised:

> From the symbolic meaning of spaces/places and the clearly gendered messages which they transmit, to straight forward exclusion by violence, spaces and places are not only themselves gendered but . . . also reflect and affect the ways in which gender is constructed and understood.

For example, the claim that women's place is in 'his home' is an old strategy that mobilises specific notions of femininity. The location of femity in private space imposed an ideology of domesticity through which females were socialised to believe that the male-privileging space called 'home' was the most appropriate space to spend their lives in, breeding and working for 'him' and 'his family' (Spain 1992). Similarly, women's places on the sportsfield were also located under specific notions of femity not only in terms of the provision of sportsfields for certain sports but also sportsgrounds themselves. Feminine and masculine space was not only segregated in terms of the sport but also the structures which were established on which to gaze at the sporting spectacle. Clearly, such segregation and exclusion also served to reinforce certain notions of gender-based identity and behaviours.

Sport as a gendered institution

The development of modern competitive sport was in part a result of women's demands for inclusion in the public sphere leading to a belief in the incipient 'feminisation' of society (Messner 1992). 'The rapid rise and expansion of organised sport during this same era can be interpreted as the creation of a homosocial institution which served to counter men's fears of feminization' (Messner 1992: 14). Sport therefore was conceived of and evolved as a masculine space, supporting male dominance not only by excluding or marginalising women, but by naturalising a connection between masculinity and the 'skills' of sport; aggression, physical strength, success in competition, and negation of the feminine (Bryson 1987).

> We can say, then, that modern sport is a 'gendered institution'. That is, it is a social institution constructed by men, largely as a response to a crisis of gender relations in the late nineteenth and early twentieth centuries. The dominant structures and values of sport came to reflect the fears and needs

of a threatened masculinity. Sport was constructed as a homosocial world, with a male-dominant division of labor which excluded women. Indeed, sport came to symbolize the masculine structure of power over women.

(Messner 1992: 16)

However, gender relations are not static. Hegemonic, or normative, notions of masculinity are constantly constructed and renegotiated within the context of evolving social structures (Kaufmann 1987). Institutions such as legal and education systems, the media, as well as the sporting arena are important sites for this negotiation. According to authors such as Messner (1992) and Kibby (1998), during the 1980s, as fewer men held positions of power in relation to the workforce and the family, they were accomplices in maintaining a hegemonic masculinity through the institution of sport both on a personal/participatory level and on a symbolic/ideological level for spectators. While this may still have arguably been the case for competitive and professional sports in North America it is noticeable that in the same period in Australia and New Zealand sport began to change within the context of evolving social structures, and from the late 1980s on was not the haven of hegemonic masculinity it once had been. Several reasons can be put forward for this shift. First, the broader acknowledgement for greater gender equity (however imperfect) in the policy settings of both countries. Second, recognition by media and advertisers of the commercial possibilities of women's sports which led to some sports, such as netball, receiving primetime television coverage, although not the same financial returns as male sports. Third, a desire by government to get greater international recognition for sporting success thereby leading to increased funding for elite men's and women's sport, particularly in the build-up to the 2000 Summer Olympics in Sydney. Fourth, the emergence of new sporting forms, such as touch football, which had mixed gender as well as single gender competitive forms, and of women's competitions in sports that had previously been considered by male dominated sporting associations as men's sports, such as cricket and rugby.

The development of touch football is arguably of substantial importance. Originally derived from various codes of football in Australia, New Zealand and Canada, the game had a strong participatory and social element which led to the development of mixed gender competition early on. Importantly, one of the reasons for the development of the game was not the non-contact nature, thereby providing from some perspectives for greater female participation, but the relatively low cost of sports equipment compared to some forms of contact football. At the same time women's soccer also started to develop a substantial player base, however, competitive adult teams have never been a feature of soccer in the same way that they are of touch football. Arguably, the introduction of touch football led to substantial negotiation and reinterpretation of male sports identity:

It was unbelievable . . . I was getting my arse kicked by a woman, she just bloody well run right passed me. I felt like a dork at first then I realised that other blokes weren't doing so well either . . . It really made me think I wasn't so special on the footy field after all.

(New Zealand male, in his 50s, reflecting on when he first played mixed touch football)

I thought it was great fun, rugby had always been a male area, where men were men and women were too . . . if you know what I mean? Anyway I was able to share my love of sport and rugby with my girlfriend, she could play and so could I without feeling guilty . . . or feeling that I was just doing a blokey thing.

(Australian male, in his late 30s)

You know it was funny, I'd been playing footy socially for a few years and often went down the park for a kick and then . . . suddenly . . . it was like *I didn't have that space to myself* anymore . . . we had to share it. It was good though, it was always the boys would play and the women would watch from the sidelines, now they were on the field as well.

(New Zealand male, in his mid-40s) (author's emphasis)

The development of touch football and its introduction to some schools also led to demands for the opportunity to participate in the contact versions of sports from which touch was derived, particularly rugby. At the same time, many of the male commercial sports organisations were seeking to broaden their spectator base for revenue purposes as well as their participatory base, particularly as this could lead to possible increases in government funding. This led to greater government and association support for sports such as women's cricket and Rugby Union, although Rugby League and Australian Rules football had only marginal growth in the development of women's competition. The development of coaching and competitions at the high school level was important for the growth of women's cricket in Australia and New Zealand while university sports clubs were significant for the development of women's rugby. However, it would also be true to say that there was substantial derision from many males regarding women's rugby: 'I'm sorry . . . but you've got to be joking, women cannot play rugby [union] they should not even be allowed on the field.' (Australian male academic and rugby coach, in his early 40s). 'Rugby is a man's game . . . women belong in the club house' (New Zealand male, late 60s).

Nevertheless, in New Zealand at least there has been relatively substantial support for the development of the women's game. This occurred in part because of a desire to encourage rugby union as a family game in order to continue to develop a player base and market given that rugby was now a professional sport. Arguably, an additional reason was that the women's rugby team were relatively more successful than the men's team in the late 1990s. This meant, for example, that the women's team were provided with the opportunity to play a test match at Eden Park in Auckland prior to a men's test and that the final of the women's world rugby cup, which

New Zealand won, was televised live in New Zealand. Similarly, the New Zealand's women's cricket test side also began to receive significant coverage in the New Zealand media. To some these measures could still be interpreted as indicating the subserviency of the women's game in the media and as mere tokenism. However, the impact on New Zealand's gender identity has been substantial in that rugby and cricket are now not just seen solely as a male sport.

> Bloody unbelievable. I never would have thought I'd live to see the day when I'd see women on the rugby field. I guess it all these bloody feminists or something. Mind you . . . I went along to see my granddaughter play and it wasn't all that bad. They played hard and real fair. There was one girl she really had a go! It makes you think though . . . I still think it's a bloke's game but hell if they enjoy it and [it] get's them into the game then why not.
>
> (New Zealand male, in his 70s)

> I couldn't believe it when the [rugby club] voted to help establish a women's team. I even went to vote against it. Some of the wives and girl-friends were real keen but. They said why shouldn't they play. Now I'm out there every weekend supporting them as much as the blokes. They're a real important part of the club. It's not like the old days but I reckon its been for the better.
>
> (New Zealand male, in his 60s)

> At first I thought I'd have to stop drinking but then I think they have a fair few drinks after the game as well. The game's changed so much. There's new rules, there's women and its professional. I tell you what though I knew things had changed when I bloody well found myself cheering the women on at that test at Eden Park.
>
> (New Zealand male, in his early 40s)

The growth of women's rugby meant changes to understanding of gendered sporting space (Law *et al.* 1999). The rugby field, especially at the club level, ceased being purely a male domain. As both males and female players used the same field for training as well as for competition. Clubroom architecture – including the provision of changing rooms and toilets – also began to change with the growth of women's rugby. The provision of male and female toilets began to become more equal. Interestingly, improved female toilet provision also began to occur in the professional sports stadiums as well. However, this tended to have occurred because to survive in a competitive entertainment market, professional rugby as the dominant stadium user needed to ensure that it attracted more female spectators. Nevertheless, the changing spectator base, along with the growth of women's rugby, has led to new considerations of what the stadium spaces mean. 'It used to be nearly nothing but blokes here. When I first started coming to the rugby it wasn't a place for women. A few came but not many. Now look at them.' (New Zealand male, in his 60s).

> I think its been really good to have women involved in the rugby. I don't have any problem with that at all. It says something about the country having grown up I guess . . . It's those toffee-nosed bastards in the corporate boxes that I can't stand.
>
> (New Zealand male, in his 40s)

Changes in sport space are also related to changed notions of regional or national identity and wider changes in society. For example, with reference to the promotion campaign by a beer brand in southern New Zealand which is also the lead sponsor of the major provincial male rugby team the following comment was made:

> This Southern Man advertising is a bit of a laugh but it's bullshit. It may promote the beer and give some funds to the Union. I mean look at crowd, some of them obviously drink the beer but there seems to be as many women here these days.
>
> (New Zealand male, in his 30s)

Another commented more at a national level:

> It used to be all men. Now we've got female trainers coming on the field. We've got female administrators and look at all the women in the ground. We've got a female Prime Minister and a woman as Governor General. The country has really changed.
>
> (New Zealand male, in his 50s)

The micropolitics of gender relations

The sum of the above comments is significant. When analysing different encounters between men and women and movements in the social space of the sportsfield or stadia, one is also focusing on the *micropolitics* of gender relations (Butler 1990, 1993). According to Butler, gender identities are the outcome of 'a process of materialization that stabilizes over time to produce the effect of boundary, fixity, and surface we call matter' (Butler 1993: 9). The norms, gazes and rituals which govern social behaviour in the stadia or the clubroom also affect the power relations between men and women and create the agenda for more or less possible gender identities (Umiker-Sebeok 1995). Through studying such geographical encounters, it may be possible to say something concerning the fluidity of the gender order and the gendered nature of space (Jackson 1991; Domosh 1997; Binnie and Valentine 1999; Longhurst 2000).

According to Burstyn (1999: 4) 'the culture of big time sports generates, reworks, and affirms an elitist, masculinist account of power and social order, an account of its own entitlement to power'. In line with many North American scholars of sports she argues that 'the ties that have bound athletes to their communities . . . are being unraveled by commercialization and free trade in athletic labour. As the ties of locality, ethnicity, and nation come more and more undone, the ties of gender, of masculinity, become increasingly important' (Burstyn 1999: 25). Yet in the antipodes such a statement does not completely ring true. Yes, the history, spaces and prevailing representations of sport in Australia and New Zealand are predominantly male, but new spaces are being opened up, and new male identities negotiated. The commercialisation of sport as in North America is not quite so rampant in Australia and New Zealand, where place relationships remain an important part of sport. Although some elements of 'Americanisation' exist it is important to recognise that in the local new possibilities for countervailing forces to universalisation of specific masculine representations exist. While speed, strength and skill remain part of the masculine identity represented in Australian and New Zealand sporting mythology, it has also become possible to represent males as emotional. The 2002 Australian Rules Grand Final at the Melbourne Cricket Ground (MCG) was a place in which the losing coach and several of his players were in tears seeking to console each other following the game. That the fact that the two coaches in the game were two of the acknowledged 'hard men' of Australian football was not lost on many (Minchin 2002). Although some did use the term 'cry babies' this particular overt display of emotion was regarded as significant as the game itself given the positive media coverage it received and the reaction from spectators.

> It was amazing. There it was in the holy of holys of Australian sport . . . and these men were crying. It was like, it really is ok to be a man and cry and show some emotion. You know? It was like all that tough footy player stuff was bullshit. You know what I mean.
>
> (Australian male, in his 30s)

Conclusions

The sportsfield is not merely a place for sporting contests: it is also a social space where gender identities are constructed. In gender studies one of the basic assumptions is the existence of a hegemonic masculinity. The dominance of particular sports-media structures by men does serve to reinforce certain mythical heroic elements of male sporting contests and particular representations of masculinity. However, these representations do have some fluidity and oppositional spaces are being created. The gendered spaces of the sportsfield and stadia should therefore be understood less as a geography imposed by patriarchal structures, and more as a social process of symbolic encoding and decoding that produces a series of homologies between the spatial, symbolic and social orders (Blunt and Rose 1994). Men are also open to transformations of their gender identities and some changes are similarly taking place with respect to the gendered nature of sporting spaces. The flow of females into what were previously male dominated sporting spaces suggests that new gender identities of sports space are being formed. Arguably, though, such shifts are more pronounced in the spectator space than on the playing field. This flow is not completely matched by a corresponding flow of men into female sporting space, though certainly in some circumstances, most notably touch football, the gendered sporting space is extremely fluid. The socio-spatial construction of gender difference establishes some spaces as primarily women's and others as primarily men's; those meanings then serve to reconstitute the power relations of gendered identity. However, since the outcome of the decoding process can never be guaranteed, contestation and renegotiation of the meaning of space is always possible (Blunt and Rose 1994). It is therefore appropriate that in the spaces associated with sporting contests and competition in Australia and New Zealand, that the notion of masculine identity and space is also being contested.

References

Andrewes, F. (1998) 'Demonstrable virility: images of masculinity in the 1956 Springbok Rugby Tour of New Zealand', *International Journal of the History of Sport* 15(2): 119–36.

Binnie, J. and Valentine, G. (1999) 'Geographical sexualities: a review of progress', *Progress in Human Geography* 23(2): 175–88.

Blunt, A. and Rose, G. (1994) *Writing Women and Space: Colonial and Postcolonial Geographies*, New York: Guilford Press.

Bond, L. (1998) 'Gender, class and urban space: public and private space in contemporary urban landscape', in Aitchison, C. and Jordan, F. (eds) *Gender, Space and Identity: Leisure, Culture and Commerce*, LSA Publication No. 63, Eastbourne: Leisure Studies Association, University of Brighton.

Bryson, L. (1987) 'Sport and the maintenance of masculine hegemony', *Women's Studies International Forum* 10: 349.

Burstyn, V. (1999) *The Rights of Men: Manhood, Politics and the Culture of Sport*, Toronto: University of Toronto Press.

Butler, J. (1990) *Gender Trouble. Feminism and the Subversion of Identity*, London: Routledge.

Butler, J. (1993) *Bodies that Matter. On the Discursive Limits of 'Sex'*, London: Routledge.

Daley, C. (1999) 'A gendered domain: leisure in Auckland, 1890–1940', in Daley, C. and Montgomerie, D. (eds) *The Gendered Kiwi*, Auckland: Auckland University Press.

Domosh, M. (1997) 'Geography and gender: the personal and the political', *Progress in Human Geography* 21(1): 81–7.

Fougere, G. (1989) 'Sport, culture and identity: the case of rugby football', in Novitz, D. and Willmott, B. (eds) *Culture and Identity in New Zealand*, Wellington: Crown Copyright.

Jackson, P. (1991) 'The cultural politics of masculinity: towards a social geography', *Transactions of the Institute of British Geographers* 16: 199–213.

Kaufmann, M. (ed.) (1987) *Beyond Patriarchy. Essays by Men on Pleasure, Power and Change*, Toronto: Oxford University Press.

Kibby, M.D. (1998) 'Nostalgia for the masculine: onward to the past in the sports films of the eighties', *Canadian Journal of Film Studies* 7(1): 16–28.

Law, R., Campbell, H. and Dolan, J. (eds) (1999) *Masculinities in Aotearoa/New Zealand*, Dunmore Press: Palmerston North.

Longhurst, R. (2000) 'Geography and gender: masculinities, male identity, and men', *Progress in Human Geography* 24(3): 439–44.

Massey, D. (1993) *Space, Place, and Gender*, Minnesota: University of Minneapolis Press.

Messner, M. (1992) *Power at Play: Sports and the Problem of Masculinity*, Boston: Beacon Press.

Minchin, L. (2002) 'No shame in grand final tears, says Matthews', *The Age*, 30 September.

Nauright, J. and Broomhall, J. (1994) 'A woman's game: the development of netball and a female sporting culture in New Zealand', *International Journal of the History of Sport* 11(3): 387–407.

Nauright, J. and Chandler, T.J.L. (eds) (1996) *Making Men: Rugby and Masculine Identity*, London: Frank Cass and Company.

Phillips, J.O.C. (1984) 'Rugby, war and the mythology of the New Zealand male', *New Zealand Journal of History* 18(2): 83–103.

Spain, D. (1992) *Gendered Spaces*, Chapel Hill and London: University of North Carolina Press.

Umiker-Sebeok, J. (1995) 'Power and the construction of gendered spaces', *International Review of Sociology/Revue Internationale de Sociologie* 6(3): 389–403.

Further reading

Binnie, J. and Valentine, G. (1999) 'Geographical sexualities: a review of progress', *Progress in Human Geography* 23(2): 175–88. The authors provide an excellent overview of the issues related to gender and space.

Longhurst, R. (2000) 'Geography and gender: masculinities, male identity, and men', *Progress in Human Geography* 24(3): 439–44. Longhurst deals specifically with male space and sexuality.

Umiker-Sebeok, J. (1995) 'Power and the construction of gendered spaces', *International Review of Sociology/Revue Internationale de Sociologie* 6(3): 389–403. Although somewhat dated, Umiker-Sebeok is extremely useful reading for those seeking an introduction to the interrelationships between space, gender and power.

Nauright, J. and Chandler, T.J.L. (eds) (1996) *Making Men: Rugby and Masculine Identity*, London: Frank Cass and Company. For a specific examination of gender and identity with respect to a specific sport (Rugby), this is the book to consult.

13

'MAN-BREASTS'

Spaces of sexual difference, fluidity and abjection

Robyn Longhurst

Summary

Men with breasts shape and are shaped by a range of complex material, discursive, and psychoanalytic spaces. In relation to material spaces many are uncomfortable at swimming pools, beaches, changing rooms and other places that require exposing their naked torso. In relation to discursive spaces men are being 'encouraged' by men's magazines and the health, fitness and beauty industries to look like GI Joe Extreme action dolls but the average weight of people in the food-secure West is increasing. Obesity is on the rise. This means rather than having toned pectoral muscles many men have soft, sagging breasts. In relation to psychoanalytic spaces breasted men disrupt understandings of sexual specificity because they are coded as feminine-fluid and as abject bodies that are subject to loathing and derision.

- My man-breasts are out of control
- Who has gynecomastia or man-breasts?
- Men's 'breasted experiences'
- Boundary disruptions: fluid masculinities and masculine fluidity
- Abjection: '*not* a good look on a man'.

My man-breasts are out of control

They look like two water balloons. They make my body look funny and lumpy . . . When I run or even walk quickly, my man-breasts jiggle like Santa's belly. Even when I'm stretched out on my back they retain a round shape . . . My back is actually fairly lean-looking and muscular. But whew, those breasts!

(Hines 2000)

This quote from Will Hines was downloaded from one of hundreds of internet sites on man-breasts, also referred to as men with breasts, man tits, beer tits, bitch tits, man boobs, and gynecomastia or overdeveloped breast tissue in men. Paul McFedries on his website defines man-breasts as 'Excess fatty tissue that causes a man's chest to resemble a woman's breasts'. The term seems to have been in reasonably common usage since the late 1990s (McFedries 2003). Some of the internet sites on man-breasts relay men's pain and shame. Some, like Will Hines's site, are humorous (men who have breasts often use humor as a way of coming to terms with their corporeality). Some sites ridicule men who have breasts. Some provide advice to those – women, men and transsexuals – who want to develop larger breasts. Some offer information and a discussion forum on male breast cancer. Some discuss male lactation. Finally, some focus on gynecomastia as a medical condition that can be 'fixed' with liposuction and plastic surgery.

While there is a wealth of information on man-breasts on the internet, the topic has been ignored by academics in masculinity studies, feminist/gender studies, and geography. Research (contemporary and historical) has been conducted on women's breasts in relation to art, self-image, surgery, breast cancer, sexuality, fetishes, nursing and maternity but not on man-breasts. In order to address this absence, in this chapter I focus on man-breasts as a cultural construction and as an important material, discursive and psychoanalytic space of masculine embodied subjectivity.

This is not a topic that I expected to write about. I am a fat[1] woman in my early 40s. My large pendulous breasts look nothing like those of the models gracing the pages of the glossy women's, or men's, magazines. They never have. I have spent a life-time reflecting on, and often worrying about, the size and shape of my breasts and not once stopped to think about breasted men. I was alerted to the issue of breasted men by participants in a research project I was conducting on fat people's constructions of subjectivity, and experiences of space and place. The research focuses mainly on the experiences of fat women, but I also interviewed four fat men in an attempt to collect a wider range of data.[2] I did not set out with the intention of asking these men specific questions about their chests-breasts but in all four interviews the topic emerged. In this chapter I draw on insights gleaned from these interviews (but do not attempt to make any generalizations from such limited data) and also on insights gleaned from many informal conversations about man-breasts with boys, men, and women in Hamilton, New Zealand.[3] Since there is little academic literature available I also draw on internet sites throughout the chapter.

The chapter begins with an examination of who is likely to have gynecomastia or man-breasts. Second, some insights are offered on the embodied experiences of breasted men and the material and discursive spaces that they shape and are shaped by. Third, is an analysis of the ways in which breasted men disrupt understandings of sexual specificity because they are coded as feminine-fluid. Fourth, I argue that not only are breasted men coded as feminine-fluid but also as abject. As abject bodies they are subject to loathing and derision. It is perhaps not surprising that some men 'chose' to have surgery to reduce their breasts. Throughout the chapter I draw mainly on Iris Young's (1990a; 1990b) research on breasts, ugly bodies and abjection, and on Elizabeth Grosz's (1994) research on sexed bodies. Neither Young nor

Grosz discuss, or even mention, man-breasts in their work, nevertheless their insights are useful for casting light on this important, but much neglected, space of masculinity.

Who has gynecomastia or man-breasts?

Gynecomastia is a medical term that comes from the Greek words for 'women-like breasts'. Though this condition is rarely talked about, it's actually quite common. Gynecomastia affects an estimated 40 to 60 percent of men. It may affect only one breast or both (Patterson 2004).

Gynecomastia, or man breasts, is especially common in teenage boys. A website entitled 'Embarrassing Problems – Breasts in Men' (see HealthPress Ltd 2004) claims that teenage boys sometimes notice that their breasts are enlarging and/or are tender. Man-breasts are related to the levels of oestrogen and testosterone being produced in the body at different times and so approximately half of all boys at some time develop breasts. 'It can start anytime after the age of about 10, and the breasts may be quite large by the age of 13 or 14. In the mid-to-late teens, they start to become smaller again, and will usually have flattened out by age 18 or 19' (see HealthPress Ltd 2004). By the end of the teen years most men no longer have enlarged breasts unless they are fat.

In fat men the breast tissue enlarges and settles. Also, fat produces oestrogen which stimulates breast development. Heavy alcohol use can also change the oestrogen-testosterone balance. Some men experience breast enlargement as a result of tumors and some as a result of old age when less testosterone is produced. Drugs are another likely reason why men develop breasts. Some drugs have an oestrogen-like effect on the breast, while others block the effect of testosterone. Some drugs mentioned on internet sites are those used to treat heart problems and psychiatric problems. Anabolic (body-building) steroids, aspartame found in many foods and drinks with artificial sweeteners, and anti-balding scalp creams are also listed as drugs that stimulate breast development in men (see HealthPress Ltd 2004). Most men develop breasts as an undesirable side-effect of taking drugs but some purchase commercially available products such as 'Addbust' with the specific aim of developing larger, fuller breasts because breasts constitute an important part of their trans-gendered/sexed subjectivity.

In short, there is a range of men who might for one reason or another at some point in their lives develop (usually involuntarily) enlarged breast tissue. Man-breasts do not just affect a few rare men who can be dismissed easily as 'freaks of nature'. Having breasts is an issue for many men and it affects in complex ways their subjectivities and relationships with place and space.

Men's 'breasted experiences'

Men's breasted bodies (like all bodies) are an interface between politics and nature, between mind and matter. They are fleshy and biological, but they are also socially and psychically constructed. Men's breasted bodies cannot be understood outside of the realms of politics, economics, sociality, culture, and the places they inhabit. Bodies and places shape each other (Nast and Pile 1998). Having breasts means different things to different men in different places. Men's breasts, like women's, vary not only in size and shape, but also in their social and cultural meanings.

It is likely, therefore, that any examination of man-breasts will be complex and exceed the space available in this chapter. My limited contribution is to begin a discussion based on the experiences of a few breasted men living in the contemporary West. Men living in other times and places will undoubtedly have different experiences. Even in the contemporary West there is a vast array of experiences amongst breasted men, for example, there are men who have breast cancer. There are also 'kiwi blokes' (see Phillips 1987 on hegemonic masculinity in Aotearoa New Zealand) who take off their shirts and yell loudly at rugby matches who seem proud of their large breasts and beer-filled bellies. It is not possible to talk about all the different constructions of breasted men.

I focus therefore only on a few men, mainly fat men, who at times feel uncomfortable about their breasts. Some men are proud of their breasts, however, most experience them as shameful. Some try to hide them by wearing thick baggy shirts and bowing their shoulders. Charles, an interviewee in his early 40s, explains:

> It's not so bad now but when I was heavier I avoided wearing T-shirts, well as much as possible. In T-shirts you can really see the saggy man-breasts. It's pretty embarrassing. I made sure I wore shirts with collars so they weren't so visible. Even now that I'm slimmer I prefer a collared shirt.

Charles was, and still is, concerned that people might notice, perhaps even stare, at his ambiguously gendered body under the soft folds of his T-shirt. This is an interesting twist on the 'evaluating gaze' on women's breasts noted by Young (1990a).

Over the past decade there has been a huge growth in media images of 'sexy men' and increasingly both heterosexual and gay men are being expected to measure up before a normalizing gaze (see Jackson, Stevenson and Brooks 2001; Mort 1996). Some men are feeling social pressure to look taut and terrific, that is, to have 'great pecs, six-pack abs and bulging biceps' (Stirling 2001: 16). Men are shaving, waxing, and using laser treatments to remove hair from their bodies to show off their lean, well-chiseled musculature to greatest effect (although some say they do it for rugby, cycling or soccer — it is easier to massage muscles, or pull off plasters if you get a graze) (Stirling 2001).

Lynda Johnston (forthcoming) discusses the 'prepped up', sculptured bodies of the 'Marching Boys' in the Auckland HERO Parade. Johnston notes that these disciplined bodies with their 'small waists, curvaceous pectoral muscles, hairless and oiled flesh are all corporeal indicators of femininity'. The firmness of these bodies, however, may be an indicator of masculinity. This is Johnston's point – these bodies transgress man-woman boundaries. New words are entering the English language such as 'himbo' (a man who is good looking but unintelligent – his female equivalent being the bimbo) and bigorexia (men who are obsessed with looking large and muscular as opposed to the anorexic who is obsessed with looking small) (see McFedries 2003).

Interviews with the four participants and subsequent informal conversations with other men indicated that boys and men are increasingly conscious of how they look and some reported avoiding activities and spaces that require exposing their torsos. In conversations a few voiced concerns about being 'too puny' but as populations in the food-secure West become on average heavier (see Langdon 1999) it is perhaps not surprising that most voiced concerns about having fat, flabby man-breasts rather than a hard, muscular chest.

For example, Phillip, aged 42, explains that he doesn't like going to swimming pools. He says:

> The one place I wouldn't go is swimming. Well, I'd probably go to the beach. That would be alright. It's more going to the pools. I've still been to them – I go with my kids but I don't feel all that comfortable there.

At the end of each interview, participants were asked to sketch themselves. Phillip drew himself with a double chin, a roll on his stomach, and man-breasts (see Figure 13.1). I asked Phillip if he would take his shirt off when he played cricket in the summer. He replied: 'No, I doubt if I would, or gardening, in my own home even, I'd still leave my shirt on'. Phillip was happy to expose his arms, legs, and feet wearing short sleeves, shorts, and sandals in the summer but not his torso.

Sam, a 55-year-old office worker whose weight fluctuates up to approximately 17 stone (or 240 lb / 109 kg), says: 'I've avoided swimming on occasions . . . 'cause of the shape of my body, being visible semi-clothed . . . In summer when you're wearing few clothes it becomes more visible'. For 22-year-old Daniel exposing his semi-clad body had been an issue since secondary school. He explains:

> Kids used to give me a hard time when I took my shirt off for swimming or PE [Physical Education] or something like that. They'd say 'Oh look you've got man-breasts'. That term, man-breasts, has been around since as long as I can remember. I think it was on 'Friends' [the television series]. It's 'cause I used to do lots of sport and it made my chest bigger but not really muscly. It didn't worry me too much 'cause I was bigger than the others and I could beat them up – just kidding [laughter]. They gave me a hard time.

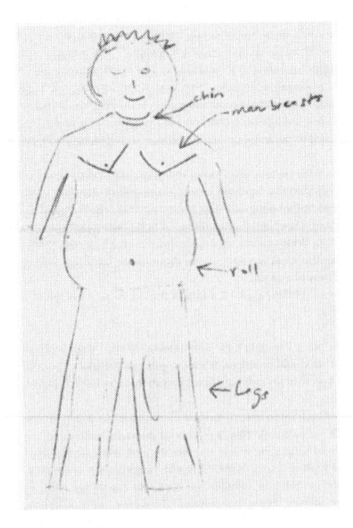

Figure 13.1 Phillip's sketch of himself (*source:* Robyn Longhurst).

Teasing boys at school about their chests/breasts seems to be common-practice in New Zealand schools. Talking to some children at a local intermediate school (children aged 11–12) in Hamilton revealed that boys who have enlarged breast tissue are described as having 'man-titties' or 'man-tits'. In an informal conversation with Brandon, a manger in his early 50s, he reported his teenage son and daughter give him a 'hard time' when he lies in his 'lazy-boy' recliner chair: They say 'Look Dad – you've got breasts!' and they laugh.

While the four participants in this research talked quite freely about sometimes feeling self-conscious about chests/breasts there were other aspects of their breasted embodiment that seemed to be unspeakable and that I felt uncomfortable asking about. The participants did not talk about sensations in their fleshy breasts, about their nipples, or about whether their breasts had become more or less of an erotic zone since growing larger. Nor did they talk about whether having breasts affected the way they viewed women's breasts, or their partner's breasts (all four of the participants had a female partner), whether they had hair on their breasts and if so how they felt about that, or whether they had any desire to bound their breasts in a bra or have a surgical reduction. Discussing these issues seemed to cross a boundary – a boundary that six foot three inch, 220 lb, 75-year-old, retired policeman, John T. Linnell (Figure 13.2), who experienced rapid and extreme male breast enlargement, was prepared to cross in publicizing his story on the internet in order to warn others about the potential dangers of the ingredient aspartame found in many foods and drinks with artificial sweeteners which he blames for his 'condition'. Linnell (1999) tells (with some humor) his story on his site:

> On 1st December 1997, at age 69, I was a normal male with average muscular chest. About noon my nipples and underlying breasts started to itch internally, not a surface itch that you can have a good scratch at. My nipples started expanding and in a couple of days I noticed my breasts seemed larger.

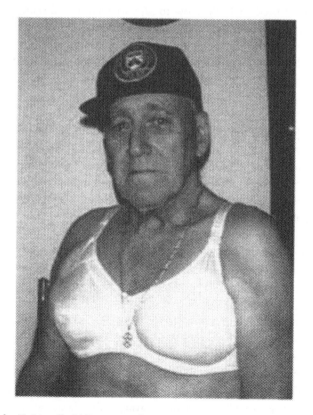

Figure 13.2 John T. Linnell, 44 E (*source:* John T. Linnell).

By 23rd Dec I was in a lot of pain from my nipples rubbing on my shirt, now about the size of a nursing mother's, and an aching feeling in my now much larger breasts . . . I went to a Sears store and bought a size 44C bra. On wearing it pain relief was almost instantaneous. My shoe size also increased from 11 to 13 in the same twenty three days. Things seemed to settle down after Christmas, the itch stopped and so did the expansion. Meantime I had been for extensive hormone checks and was warned that if the prolactin levels went up I might just become a breast feeding daddy . . . For all of 98 I wore a 44C which I filled . . . Come 1 Dec 98 and the itch started again. By Christmas the C size was too tight and I tried . . . a DD. It fitted beautifully. In Dec 2001 I had to get a 44E bra over the Internet since local stores carried no larger than DD.

Now today, 28 Nov 2003, my shoes again being uncomfortable I got my feet checked and found I need a size 16 shoe, feet still growing at age 75. My 44E bra is getting full and I will have to get an EE shortly, and so here I am sitting in my underpants and bra writing the story. From the occasional itch I know new expansion is going on. Breast reduction surgery is the only way to be rid of them, but not much sense when they are still growing. I guess when I need a wheelbarrow to support them I shall reluctantly have to let some butcher have a few pounds of hamburger.

Linnell explains that there is no way to hide his breasts, so for comfort he wears a T-shirt most of the time, except when he needs to dress formally. The responses to his embodiment are varied. He says:

To date I have only had one unpleasant experience. In the summer of 1998 I was downtown and a couple of yahoos yelled f———ng faggot at me as they approached. Their grins froze on their faces as I picked them both up by their shirt fronts, looked at them and asked if they wanted to make something of it and then threw them on their backsides on the sidewalk. Not something they expected from a 68 year old.

The assumption made by abusive 'yahoos' that Linnell was a 'faggot' indicates that some people are very uncomfortable when confronted with an ambiguously gendered body. Men who have breasts disrupt the border between heterosexuality and homosexuality, and between masculinity and femininity.

Elizabeth Grosz (1994: 203) notes 'there are particular bodily zones that serve to emphasize both women's difference from and otherness to men'. These zones are culturally determined rather than 'natural' or 'biological'. One such zone is the breasts. Sally Munt (1998) offers an insightful account of being a 'butch' in the 'Ladies toilet' and the boundary disruptions this causes (see also Kath Browne's chapter in this volume). Appearing masculine, including having a flat chest, can result in women being abused. In this case, having visible breasts resulted in a man being abused.

Boundary disruptions: fluid masculinities and masculine fluidity

In the West 'Breasts are the symbol of feminine sexuality' (Young 1990a: 190). Men's breasted bodies, therefore, tend to make little sense. For women 'breasts are an important component of body self-image' (Young 1990a: 189) but what about when breasts are an important component of body self-image for men? In this section I argue that man-breasts cause considerable cultural anxiety not only because they disrupt understandings of sexual specificity, but also because they are coded as fluid rather than as solid. Men's breasts befuddle normative understanding of what it means to *have* breasts (or rather to *be* breasts – they are part of us rather than objects we possess) in a highly masculinist culture.

Iris Young (1990a) argues that Western culture fetishizes (women's) breasts. A fetish, she explains 'is an object that stands in for the phallus' (p. 190). Young elucidates: 'The "best" breasts are like the phallus: high, hard, and pointy.' (p. 190). The ideal breasts, she says, look like Barbie's. They sit high on the chest rather than sag, are firm-looking, and are large but not too bulging. As Young points out, this norm is contradictory because when breasts are large, gravity tends to pull them down and they are likely to be soft and fleshy rather than firm. Most women's breasts do not conform to the 'ideal' breasts depicted in the dominant media (see Ayalah and Weinstock's 1979 collection of photographs of the breasts, and accompanying accounts, of fifty women). Exercise, creams, specific items of clothing, bras and surgery are all used as means to make the breasts appear firm and high. In other words, they are used to solidify the breasts. Firm, contained breasts are privileged over soft, fluid breasts, but why?

Luce Irigaray (1985) argues that in Western philosophical models of ontology, fluids, fluidity, and the viscous are subordinated to that which is concrete and solid. Also, fluids, fluidity, and the viscous are implicitly associated with femininity, maternity, pregnancy, menstruation and the body. Solidity is associated with masculinity and rationality. 'Solid mechanics and rationality have maintained a relationship of very long standing, one against which fluids have never stopped arguing' (Irigaray 1985: 113). Irigaray is not necessarily proposing that women's bodies are inherently fluid while men's bodies are solid, but that men's and women's bodies become coded in this way. Grosz (1994: 203) explains 'women's corporeality is inscribed as a mode of seepage'. She suggests that women have the same degree of solidity as men but 'they are represented and live themselves as seepage, liquidity' (Grosz 1994: 203; also see Longhurst 2001: 9–32).

These insights into a politics of fluidity/solidity are useful for understanding the 'breasted experiences' (Young 1990a) of both women and men. In relation to women, Young (1990a: 195) argues: 'Without a bra, the fluid being of breasts is more apparent . . . Many women's breasts are much more like a fluid than a solid; in movement, they sway, jiggle, bounce, ripple even when the movement is small.' Wearing a bra solidifies the breasts or, as Susan Bordo (1993: 20) puts it, wearing a bra 'disciplines' the breasts.[4] Breasts must be normalized – lifted and curved – to approximate the ideal breast (Young 1990a: 195).

This unacceptability of 'jiggly' breasts poses difficulties for some women who, for whatever reason, do not like to wear a bra, but it also poses difficulties for breasted men. It is unacceptable for men to have soft, fleshy breasts but it is equally, possibly even more, unacceptable for men to attempt to solidify their breasts by wearing a bra. Unlike women, men are unable to 'bolt down' and discipline their breasts thus further encoding them as fluid rather than as solid.

Daniel notes that when he walks fast or runs he feels 'a bit self-conscious' because his breasts bounce up and down. He doesn't like how this feels and he doesn't like others looking at him. It discourages him from exercising. Some men given an opportunity would, more than likely, wear a bra in order to constrain, firm up, solidify the softness and indeterminacy of their breasts but currently for most men wearing a bra is out of the question because it would remove them still further from spaces of hegemonic masculinity.

When the aforementioned John Linnell bought his first bra (44C) he pretended it was for his wife. When he bought the second bra (44DD) there was no more pretending – he 'told the sales lady up front' that it was for him. Given the association between femininity and fluidity, and masculinity and solidity, coupled with the difficulties men have fitting, buying and wearing a bra, breasted men occupy a difficult and transgressive space.

Breasts are not only fluid in relation to their shape, but also as Grosz (1994: 207) notes, they are sites of (potential) flow such as colostrum (the earliest form of milk that women can produce at approximately 28 weeks pregnant) and milk. Breasts are thought to embody maternity. During pregnancy and after the birth of an infant most women experience an enlarging of their breasts. But like the transsexual, the breasted man remains outside women's lived experiences of maternity. Grosz (1994: 207–8) explains: 'The transsexual may look like a woman but can never feel like or be a woman. The one sex, whether male or female or some other term, can only experience, live, according to (and hopefully in excess of) the cultural significations of the sexually specific body.' Grosz points out that there is an irreducible difference, a gulf, between the sexes. Rather than respect this gulf masculinist regimes disavow it – 'these regimes make the other over into a (lesser) version of the same' (Grosz 1994: 208).

The man with breasts is seen as a lesser version of a man, and a lesser version of a woman. It is not possible for him to be seen as anything but a poor copy of both. There is no space for him to exert his difference and he suffers the violence that attempts to efface his difference. The breasted man, therefore, is likely to attempt to cover up or even to get rid of his breasts. Some men resist phallocentric constructions of sexed/gendered bodies and take pleasure in their breasts but most find the spaces of ambiguity that they are forced to occupy as 'ugly' (Young 1990b), 'abject' (Kristeva 1982) others, difficult.

Abjection: '*not* a good look on a man'

When a friend asked what I was currently researching and I replied 'man-breasts' she screwed up her face and exclaimed 'Oh yuck – breasts are *not* a good look on a man!' Man-breasts tend to be represented as abject. Kristeva (1982) argues that the abject provokes fear and disgust because it exposes the border between self and other. This border is fragile. The abject threatens to dissolve the subject by dissolving the border. The abject is also fascinating, however; it is as though it draws the subject in order to repel it. Young (1990b: 145) uses the notion of the abject to argue that some groups become constructed as 'ugly or fearsome'. She states: 'Racism, sexism, homophobia, ageism and ableism are partly structured by abjection, an involuntary, unconscious judgment of ugliness and loathing' (Young 1990b: 145). 'Fat oppression' is also partly structured by abjection. Many men have breasts because they are fat. Brown (1989: 1) defines 'fat oppression' as:

> hatred and discrimination against fat people . . . solely because of their body size. It is the stigmatization of being fat, the terror of fat . . . It is the equation of fat with being out-of-control, with laziness, with deeply-rooted pathology, with ugliness.

Men with large bellies and breasts are often constructed, both by themselves and by others, as 'despised, ugly, and fearful bodies' (Young 1990b: 142). Sometimes they are even constructed as grotesque. Consider, for example, the character 'Fat Bastard' played by Mike Myers in the movie 'Goldmember'. Fat Bastard has huge rounded hairy shoulders, a large belly and ample womanly breasts. He is supposed to personify grotesqueness and functions as the alter-ego to his suave, spy counterpart, Austin Powers. Fat Bastard has become known for delivering (in a Scottish accent) lines such as 'Get in my belly!' 'I'm dead sexy!' and 'Frisky are we?' Man breasts have also been represented in a number of television comedies including 'Seinfeld' and 'Friends'.

Given such representations of fatness and man-breasts as grotesque and/or funny it is perhaps not surprising that some men who are fat, and/or have breasts, decide to undergo surgery. 'The procedure removes fat and or glandular tissue from the breasts, and in extreme cases removes excess skin, resulting in a chest that is flatter, and firmer, and better contoured [read: more masculine]' (Patterson 2004).

Fat men tend to be advised to attempt to lose weight through exercise and diet before resorting to surgery but some proceed with surgery. According to a report on 'male vanity' in New Zealand (Stirling 2001: 20) 'Increasingly, men are seeking breast reductions (the breast tissue can sometimes even be the result of steroids).' (But, still far more women go under the surgeon's knife for breast enlargements, lifts and reductions each year – see Davis 1995 and Yalom 1997 on cosmetic surgery.)

175

Conclusion

This chapter has examined an 'embodied space' of masculinity – man-breasts. This is a topic that has been much neglected in masculinity studies, feminist/gender studies and geography. Men with breasts shape and are shaped by a range of complex material, discursive, and psychoanalytic spaces. In relation to material spaces many are uncomfortable at swimming pools, beaches, changing rooms and other places that require them to expose their naked torso. In relation to discursive spaces men are being 'encouraged' by men's magazines and the health, fitness and beauty industries to look like 'GI Joe Extreme action dolls' (Stirling 2001) but the average weight of people in the West is increasing. Obesity is on the rise (Langdon 1999). In relation to psychoanalytic spaces breasted men disrupt understandings of sexual specificity because they are coded as feminine-fluid and as abject bodies that are subject to loathing and derision.

Having reached the end of this chapter I feel that I still have more questions than answers. The only thing I am certain about is that the experiences of breasted men and the spaces they occupy is a much neglected topic in academic research. My hope is that in the future we will see more researchers address the topic of breasts, drawing on non-essentialist understanding of sex, gender, bodies and spaces.

Notes

1 According to the *Collins English Dictionary* (1979: 529) the term 'fat' refers to 'corpulence, obesity, or plumpness'. In the contemporary West calling someone 'fat' is likely to cause offence because 'Fat is seen as a sign of moral and physical decay. Fat people are stereotypically constructed as undisciplined, self-indulgent, unhealthy, lazy, untrustworthy, unwilling and non-conforming' (Bell and Valentine 1997: 36). However, individuals and groups such as the National Association to Advance Fat Acceptance (NAFFA) in the US have reclaimed the word 'fat' from its use as derogatory slang to be used as a positive signifier by those wanting to end discrimination based on body size (Howells 1993: 19). It is in this spirit of reclaiming that I use the term 'fat' in this chapter.

2 The four participants were aged between 22 and early 50s. The participants all identified as Pakeha (New Zealand European), and as heterosexual. The interviews lasted between 30 and 75 minutes and were transcribed. I did not ask participants their weight and height partly because I did not want to be insensitive and partly because being fat is not simply about being a particular physical size or having a particular 'body mass index', it is also about what Colls (2002: 219) refers to as 'emotional size'.

3 Hamilton is located in the North Island, approximately 100 km South of Auckland (New Zealand's largest city), and has a population of 166,128 people (Statistics New Zealand 2003). It is suburban in character and surrounded by land used mainly for diary production.

4 Young (1990a: 195–6) also notes that 'without a bra, the nipples show . . . Nipples are no-nos, for they show the breasts to be active and independent zones of sensitivity and eroticism'.

References

Ayalah, D. and Weinstock, I. (1979) *Breasts: Women Speak About Their Breasts*, New York: Summit Books, Simon and Schuster.

Bell, D. and Valentine, G. (1997) *Consuming Geographies: We Are Where We Eat*, Routledge: London.

Bordo, S. (1993) *Unbearable Weight: Feminism, Western Culture and the Body*, Berkeley: University of California Press.

Brown, L. (1989) 'Fat-oppressive attitudes and the feminist therapist: directions for change', *Women and Therapy: A Feminist Quarterly* 8(3): 19–30.

Collins English Dictionary (1979) London and Glasgow: Collins.

Colls, R. (2002) 'Review of "Bodies out of bounds: fatness and transgression"' in Braziel, J.E. and Lebesco, K. (eds) *Gender, Place and Culture* 9(2): 218–20.

Davis, K. (1995) *Reshaping the Female Body: The Dilemma of Cosmetic Surgery*, New York: Routledge.

Grosz, E. (1994) *Volatile Bodies: Towards a Corporeal Feminism*, Indianapolis: Indiana University Press.

HealthPress Ltd (2004) 'Embarrassing problems.com. Breasts in men', Online. Available at http://www.embarrassingproblems.com/pages2/breasts_g.htm (accessed 19 February 2004).

Hines, W. (2000) 'Man breasts are out of control', Online. Available at http://ibiblio.org/spite/bile/mb/ (accessed 19 February 2004).

Howells, S. (1993) 'A fat lot of good', *Listener*, 8 May, 138(2770): 18–21.

Irigaray, L. (1985) *This Sex Which is Not One* (trans. C. Porter with C. Burke), New York: Cornell University Press.

Jackson, P., Stevenson, N. and Brooks, K. (2001) *Making Sense of Men's Magazines*, Cambridge: Polity Press.

Johnston, L. (forthcoming) 'Man – Woman', in Cloke, P. and Johnston, R. (eds) *Deconstructing Geography's Binaries*, London: Sage Publications.

Kristeva, J. (1982) *Powers of Horror: An Essay on Abjection* (trans. Leon S. Roudiez), New York: Columbia University Press.

Langdon, C. (1999) 'We're all getting too fat, survey shows', *The Dominion*, 26 August, Edition 2: 3.

Linnell, J.T. (1999) 'Victims reports and recovery stories', Online. Available at http://www.aspartame.ca/page_a9.html (accessed 19 February 2004).

Longhurst, R. (2001) *Bodies: Exploring Fluid Boundaries*, London and New York: Routledge.

McFedries, P. (2003) 'The word spy – man breasts', Online. Available at http://www.wordspy.com/words/manbreasts.asp (accessed 19 February 2004).

Mort, F. (1996) *Cultures of Consumption: Masculinities and Social Space in Late Twentieth-Century Britain*, London and New York: Routledge.

Munt, S.R. (1998) *Heroic Desire: Lesbian Identity and Cultural Space*, London: Cassell.

Nast, H. and Pile, S. (eds) (1998) *Places Through the Body*, London: Routledge.

Patterson, R.S. (2004) 'Gynecomastia – correction of enlarged male breasts', Online. Available at http://plasticsurgerycan.com/breast-reduction.htm (accessed 19 February 2004).

Phillips, J. (1987) *A Man's Country? The Image of the Pakeha Male – A History*, Auckland: Penguin Books.

Statistics New Zealand (2003) *New Zealand in Profile*, Wellington: Statistics New Zealand, Te Tari Tatau.

Stirling, P. (2001) 'Does my chest look big in this?' *New Zealand Listener*, 20 October, pp. 16–21.

Yalom, M. (1997) *A History of the Breast*, New York: Alfred A. Knopf.

Young, I. (1990a) *Throwing like a Girl and other Essays in Feminist Philosophy and Social Thought*, Indianapolis: Indiana University Press.

Young, I. (1990b) 'The scaling of bodies and the politics of identity', in *Justice and the Politics of Difference*, Princeton: Princeton University Press.

Further reading

Worth, H., Paris, A. and Allen, L. (eds) (2002) *The Life of Brian: Masculinities, Sexualities and Health in New Zealand*, Dunedin, University of Otago Press. This book begins with an introduction in which the editors define the term 'masculinities' and discuss various approaches to studying masculinities. Drawing on feminist poststructuralist theories on the body they argue that the 'embodiment of masculinity is specific to the culture and historical moment to which it belongs' (12). The book contains ten essays on a variety of topics including 'Tits as an Accessory for a Maori/Pacific Queen', 'Gay-Disabled Masculinity' and '(Hetero)sexual Pleasure'.

Law, R., Campbell, H. and Dolan, J. (1999) *Masculinities in Aotearoa/New Zealand*, Palmerston North, Dunmore Press. This book also provides a useful insight into the ways in which various masculinities are constructed in New Zealand. Most of the 13 contributors, from a variety of disciplines, have linked their chapter in one way or another to Bob Connell's recent work on masculinities. It is a valuable resource for anyone, but especially for undergraduates, with an interest in issues of power, identity, gender, sexuality, and space.

Young, I. (1990) 'Breasted experience: the look and the feeling', in her book *Throwing Like a Girl and Other Essays in Feminist Philosophy and Social Thought*, Indianapolis: Indiana University Press. In this fascinating and rare essay Young adopts a feminist perspective to discuss 'Breasts as Objects', the 'Woman-centered Meaning' of breasts, 'Motherhood and Sexuality', and 'The Knife at the Breast'. She does not discuss the experiences of breasted men, nevertheless, her focus on gendered embodied identities means that the essay provides a useful starting place for anyone who wants to examine the issue of 'Man-breasts'.

14

TATTOOS IN PRISON

Men and their pictures on the edge of society

Janine Janssen

Summary

In European history, tattoos have been used in order to brand slaves or delinquents. Nowadays, an impressive collection of tattoos can be found amongst inmates in penitentiaries all over the world. Rather than forced upon, though, these tattoos have been set on a voluntary basis largely for the purpose of inscribing personal or/and group identity. As such prison tattoos have become expressions of power, empowerment and control.

- Prison as a space
- Prison, space and gender
- Tattoos as texts
- Tattoos in prison.

Introduction

Every (sub)culture tends to have its own ideas about the body and bodily inscriptions such as tattoos. In the workplace of a bank office it is inappropriate for a male employee to publicly display the tattoos on his chest. The same person might not be regarded as confrontational if his workplace was located in a harbour. These ideas about meaning and acceptability of bodily inscriptions develop within social and cultural contexts.

There is, therefore, a relationship between attitudes towards tattooing and the space in which people decide to sport tattoos. In this chapter, I shall examine the meaning of tattoos that are worn by male inmates in the space of prisons. I will begin by exploring the relationship between prison, gender and space. What kind of spaces do prisons consist of and to what extent are they gendered? I shall then provide a brief historical overview of the history of the art of tattooing in the Western world. Two specific examples of tattooing in prison, one from Russia and the other from the United States, are used to illustrate issues of control and identity/ies.

179

Prison as a space

The first thing that comes to mind when discussing prison space is its sheer physical reality: a prison is a building that makes up part of the urban landscape or, more likely, a remote rural area. This is a rather static and absolute way of referring to prison space. However, when understood in a more relative manner, this space is an integral part of socio-economic organization, processes and relations whereby it is 'the medium, as well as the outcome, of social action' (Rose *et al.* 1997: 7). The correctional system is a medium used by society for dealing with delinquents and felons. The outcome of a need to deal with such individuals is the construction of and confinement to prisons. As such, prisons emphasize the relation between the social position of an individual and his or her location in space (McDowell 1996). Finally, the concept of space may also be used as a metaphor (Rose *et al.* 1997). We speak of 'sentencing someone to prison'. The physical reality – the building – is used to describe the abstract idea of punishment.

Inside prison something very unique with regard to space occurs: all aspects of life take place in the same space and under the same command. This is a distinguishing feature of the so-called 'total institution', a concept that was introduced by the sociologist Goffman in the 1950s. According to Goffman, this type of institution includes its members in a totalitarian way. Another feature of a total institution is that daily activities are carried out in the company of a lot of other inmates or inhabitants, who are all treated in an equal way. These daily activities are planned by the staff, they are often obligatory and they primarily serve institutional purposes (Goffman 1977).

There is, therefore, little spatial mobility within these institutions, since everything happens within the same structure: working, sleeping, eating, and other activities all take place within the boundaries of the prison. Within prisons, there is a special space for each of these functions. Where an inmate carries out each activity is decided by the staff, but is also proscribed by the structure of the building. Most of the 'sub-spaces' (e.g. the cell, kitchen, work area or showers) within the institution are designed for just one or a few limited functions only. The individual cell is for sleeping in and passing spare time, but not for receiving visitors. When the weather is fine, an inmate cannot simply decide on his own accord to enjoy his morning coffee in the outdoor courtyard. Consequently, an inmate has limited opportunities for physical mobility within the prison. Nowadays, an inmate's physical mobility is not only controlled by high walls, bars and gates. Technological changes have also entered the space of prisons: cameras and other electronic security devices have become an (in)visible part of the penitentiary.

Inside the institution, space is also linked to social relations. Being in prison has consequences for the relative space of inmates: an inmate is not only physically removed from society, but his social relations with (significant) others are also affected given that the inmate is limited in the performance in his ordinary social roles (e.g. being a parent, an employee or spouse). These social functions influence the way in which an individual perceives his personal identity: 'I am a single man, a student, a father of three' and so on. Yet in the eyes of the institution, an inmate is first and foremost an inmate. In this way, within the space of a prison, the inmate is largely stripped of his identity as he himself defines it.

In his groundbreaking criminological study 'The defences of the weak' (1972), Mathiesen states that inmates in the space of prison tend to separate life inside the institution from that on the outside:

> The separation between inside and outside is reflected in a series of state-
> ments and expressions among inmates . . . For example, inmates' extremely
> frequent use of the word 'society' regularly implies that they are distant
> from it ('How can we get back to society?' 'Society must do something').
>
> (Mathiesen 1972: 76)

Mathiesen observes that the inmates can use space outside as a 'comparison refer-ence-point'. He explains that imprisonment implies the stigmatization of the inmate: 'It may be said with considerable certainty that [inmates] belong to . . . the very bottom floor of outside society in terms of status' (Mathiesen, 1972: 73). The outside can also be perceived in more positive terms. Outside of prison there are pleasures, but due to his detention, the inmate is deprived from participating in pleasurable activities. In contrast, some inmates use the space inside the institution as a comparative point of reference. In so doing, the inmate does not have to face the (frequent) failures, stigmatization and frustrations that he may encounter outside in society. In other words, what happens inside the institution can become more meaningful and important to its inmates.

Prison, space and gender

In the following, the relationship between gender and space within the context of prison will be explored. First, the relation between gender and space must be con-sidered. A popular dichotomy has been between public and private space. This dis-tinction is gendered: men have generally been associated with the realm of public space, where they are frequently considered to behave in a loud and aggressive fashion (Morgan 1993). Women, on the other hand, have often been associated with the private sphere: the household, care, passion, reproduction and unwaged labour (Duncan 1996).

However, not all space is clearly private or public. A good example of this is the penitentiary. Since the general public has no access to it (like they would have to a shopping mall), prison cannot be considered a public space. Yet it is also difficult to state that prison is a purely private space. Although the inmates live in the institution and may call it 'home'[1] during their stay behind bars, prison is also a place of work for guards, psychologists and other officials employed there. One may, perhaps, call the cellblock the private space of inmates and the working areas and the courtyard the public space. Nevertheless, within the context of prison, it is rather difficult to see these spaces as being gendered as was described above. For it is only individuals of one sex alone who are detained in a specific penal institution.[2]

Goffman focused on aspects of the total institution in order to understand how inmates function in this type of organization. He distinguished different ways for inmates to deal with the pressure of being institutionalized: they can withdraw into their own world or they can reject everything related to the institution. Another way is the 'colonization' of inmates: they start to feel at home in the institution and when they compare life outside to life inside, they come to the conclusion that living in the institution is not so bad at all. It is also possible that inmates 'convert' themselves to the beliefs and the goals of the staff and the institution. A 'convert' usually tries to become a model inmate (Goffman 1977). In the literature on penitentiaries, Goffman's approach, and that of other researchers who tend to stress that the way inmates behave is a reaction towards the deprivation caused by the institution, is labelled the 'deprivation-model'. The deprivation-model has, however, been criticized by scholars stating that the personal background (socio-economic position and criminal background) also influences an inmate's behaviour. This model is called the 'importation-model'. In research, both models are used as complementary to one another because aspects related to the organization of the institution as well as features of the personal background of inmates are assumed to be of importance in trying to understand the conduct of inmates (Thomas 1977; Thomas et al. 1978).

The way in which inmates behave in the space of prison is influenced by their personal background. Aspects such as ethnicity, social class, prior experiences in detention, criminal record, relationships with family and significant others, influence the way in which detention is experienced. A first-time offender may be deeply affected by his first encounter with the space behind bars. An inmate who has already done time might experience detention simply as a routine. Another important element of that background is gender. There is a substantial body of literature in criminology (and other disciplines) on how inmates cope with the pressures of detention. Reactions described vary from being very passive to becoming openly resistant to the prison-system. Most of these latter publications, however, focus exclusively on men. Studies of female prisoners tend to emphasize the passivity of women (Bhavnani and Davis 1995). In general one could state that female inmates are more often described as 'model inmates', whereas (aggressive) forms of open resistance are usually ascribed to males.

Tattoos as texts

A glimpse at the history of tattooing in the Western world

The Western history of tattooing is a rather complex one. Although I cannot do justice to this complexity here, I will briefly present a few 'highlights' of this history. In 1774, Captain James Cook returned from one of his voyages of discovery in the

Pacific, bringing with him a tattooed native. Although Cook introduced the word 'tattoo'[3] to Europe, the procedure was already known in the 'Old World'.[4] During the Middle Ages and until the eighteenth century, delinquents received marks on their skin through burning or 'pricking' (Schuler 1997; DeMello 2000). In Cook's time, this form of stigmatization and identification was also applied to slaves (Schuler 1997).

From the seventeenth to the twentieth century, tattooed natives from islands in the Pacific were displayed in Europe and North America. In the beginning of the twentieth century, it was more frequently tattooed westerners who were put on show at fairs.[5] These people were presented along with adventurous biographies (DeMello 2000; Schuler 1997). The tattoo demonstrated an individual life history and was thus a reflection of one's personal identity. Especially those who thought of themselves as outsiders chose to wear tattoos in order to demonstrate their weak links with mainstream society. This marked a sharp contrast to the use and meaning of tattoos in traditional tribal societies where tattoos were considered to be a 'ticket in'. Tribal tattoos represented the relationship between an individual and the social and religious structure of the tribe (Schuler 1997). In the West, however, more attention was eventually paid to the sensational aspects of tattooing. In the nineteenth century, one of the 'founding fathers' of criminology, the Italian Cesare Lombroso, began to focus on the tattoos sported by criminals. Among other things, Lombroso spoke about the alleged propensity of criminals to get multiple tattoos on (sensitive) parts of the body. This celebrated criminologist thought that criminals shared a lesser sensitivity to pain with 'savages' (Caplan 2000). Consequently other criminologists, such as Lacassagne, also embraced this particular theme (Klees-Wambach 1976).

Until 1890, tattoos were produced by hand, using methods from the Polynesians. In 1891 the tattoo machine was patented. Tattoos made with a machine were cheaper, less painful and easier and faster to administer. This contributed to the spread of tattooing among males throughout the lower classes. In the twentieth century, the art of tattooing developed into a business (DeMello 2000). Although in the Western world the tattoo has often been used as an instrument for stigmatizing individuals and pushing them to the edge of society, it has been repeatedly stated that the application of the tattoo 'is rapidly losing its deviant status' (Vail 1999: 271). The tattoo is no longer only or mainly employed to demonstrate the rejection of mainstream society. To some extent, the tattoo has now also become an element of fashion. Nowadays, the spread of the application of tattoos also seems to be more 'democratized', meaning that wearers of tattoos can be found in all social strata. Today different styles and symbols can be combined and one can even get temporary tattoos that can be washed off. It is perhaps too simple to present the modern application of tattoos as being just superficial fashion statements (Schuler 1997) as (Western) people, delinquents and non-delinquents, continually search for ways of expressing their individuality. Deciding to start to wear (even temporary) tattoos must still be considered a way of expressing a personal identity.

Tattoos and gender

> Clothing, jewelery, makeup, cars, living spaces, and work all function to
> mark the subject's body as deeply as any surgical incision, binding indi-
> viduals to systems of significance in which they become signs to be read (by
> others and themselves) ... Bodies speak, without necessarily talking,
> because they become coded with and as signs.
>
> (Grosz 1995: 35)

Grosz sees the body as a 'surface of inscription'. To the above mentioned list
of 'inscriptions' (like clothing and jewellery) tattoos can be added because these
are literally inscriptions that make the body speak. But the human body is gendered,
therefore there is a link between gender and the way in which bodies are inscribed
(Grosz 1995). Or in Grosz's words: 'These [sexual] differences ensure that even
if the text to be written seems the "same" one, the body's positive contribution
to the "text" produced ensures that the inscribed message will be different'
(1995: 36).

Tattooist Samuel Steward wrote a fascinating account of tattooing in the 1950s
(Steward 1990). He described how, in those days, he refused female customers
unless they were over 21, married, accompanied by their husbands and, moreover,
had written proof of their marriage with them. According to Steward, 'nice
girls don't get tattoos'. 'Nice girls' were attractive, heterosexual women with a
middle class background. Women who did not fit into this specific category,
for example lesbians, could get tattoos (Steward 1990). According to DeMello
(2000), women were largely absent from the Western tattoo scene until the
1970s. DeMello quotes the tattoo-artist Lyle Tuttle: 'Women's liberation came
along ... So then women started getting tattooed' (DeMello 2000: 75). Although
people generally do not raise their eyebrows if a tattooed female passes by, tattoos
are still more closely associated with men than women (Mascia-Lees and Sharpe
1992).

As described above, the tattoo has become increasingly mainstream, but it is still
groups of males that are traditionally associated with tattoos, for example, sailors,
soldiers, bikers and inmates. These groups have in common that their members live
adventurous and even 'exotic' lifestyles and engage in typical male activities (like
travelling and working all over the world, going to war, etc.). Another similarity is
that these groups can also be found in total institutions, such as the army, navy and
penitentiaries. It seems that for these particular groups of men sporting a tattoo is
not only a form of establishing an identity (e.g. as a sailor or a biker), there is also a
relation between tattoos and male bonding. By wearing a specific tattoo they can
show each other and the rest of the world what kind of men they are (e.g. gang
members or soldiers). Not only the final result – the tattoo – but also the process of
'inscribing' the body can be a manner for expressing one's masculinity. Tattooing
each other or visiting a tattoo parlour as a group can further strengthen the process

of male bonding, even of becoming a brotherhood. Within these particular male groups the painful application of tattoos can also function as a 'rite of passage'. In order to prove that one is worthy to obtain membership, an individual male must be able to stand the pain. The element of pain can also be used the other way around as a form of punishment. It is known – as I will discuss below – that in some prisons, inmates tattoo fellow detainees against their will. The attributed pain is then not a mode for demonstrating one's masculinity, but for one's lack of it, because the masculine identity is (partially) formed by being in control of one's own body (Morgan 1993).

Tattoos in prison

In spite of the discussion above, popular perception continues to connect tattoos with deviant subcultures[6] and the space of penitentiaries. After looking into different studies on (prison) tattoos, sociologist Clinton Sanders (1989) argued that different researchers regard tattoos as an indication of a penchant for violence, a tendency towards self-destructive behaviour, a pathological need for attention or a tendency to engage in certain forms of property crime. Many analysts regard the tattoo as a means of defending oneself against the 'identity-stripping' by the total institution (Sanders 1989: 40). In other words: 'A prisoner has nothing of his own, not even decent clothes . . . Because the only thing that belongs to a prisoner is his body, it can be violated or bartered or turned into a picture gallery' (Bronnikov 1993: 58). It is interesting to note that not all authors on tattoos in prison distinguish between tattoos that are made in penitentiaries and those that have been made outside the prison walls. In penal institutions inmates from lower strata of society are usually overrepresented (Mathiesen 1965; Reiman 1995). Therefore, these inmates are likely 'to import' the bodily decorations that are representative of their low social status outside.

In Europe, the tradition of prisoners tattooing themselves and each other dates back at least three hundred years (DeMello 2000). However, in the Netherlands – where I conducted research in penitentiaries during the last decade[7] – little is known about the tattoos of inmates. One of the respondents to my recent study on prison tattoos, a member of the board of directors of the penitentiary Ter Apel in the north of the Netherlands, told me that:

> In the Netherlands tattooing in prison is not prohibited by law. But that does not necessarily mean that the staff agrees with this practice. One of our biggest concerns is that prisoners will infect each other with AIDS by tattooing with dirty needles.

During my research in Dutch prisons, I often noticed that many inmates sported tattoos. Since I was unable to find more information about prison tattoos in Dutch penological literature, I examined the bodily decorations of sixty imprisoned robbers. Unfortunately, I did not have the opportunity to speak to these inmates in person about their tattoos. Instead, I used descriptions from Dutch police records.[8] Of these robbers, 25 had tattoos, mostly situated on their arms. The most popular were tattoos of animals (e.g. eagles, dragons and tigers) and short texts or words (like 'mother', 'fuck off' or the name of a woman). Furthermore, these inmates had a lot of symbolic motives on them (e.g. tribal, celtic symbols and hearts). Although the Dutch police[9] register descriptions of tattoos and other bodily decorations (e.g. piercings) and their location on the body in their official records, many important questions about the meaning of these for delinquents and their relationships and social status remain unanswered. In order to shed light on the above issues, I will use cases from the Russian and American context as tattooing behind bars is well documented in these countries. It must be emphasized though that the living conditions in Dutch prisons and the relatively liberal Dutch penal policy in general cannot be compared to the Russian or American national prison systems. Nonetheless, insights might be gained that could stimulate further research in the social and cultural context of Dutch prisons.

Tattoos in Russian prisons

At the end of the nineteenth and the beginning of the twentieth century, Russian criminologists and other social scientists started to study the life of inmates and those that had been exiled to Siberia more closely. It was noticed that the world of inmates was extremely hierarchical and that it had its own social practices and culture. A special element of this culture was that inmates tattooed their own bodies (Schrader 2000).

According to Schrader, Soviet and Russian convicts continued to use tattoos in order to stress the solidarity of the group and as a rite of passage throughout the twentieth century. Tattoos were also part of a secret language that excluded the authorities (Schrader 2000). Arkady G. Bronnikov's work, a collection of more than 20,000 photographs of tattooed inmates, assembled while working in Russian prisons,[10] illustrates Schrader's findings and sheds more light on tattooed codes and social relations in prison.

Bronnikov's tattoo collection

Inmates told Bronnikov that 'in most cases . . . they started wearing tattoos only after they had committed a crime. As convictions increase and the terms of incarceration become more severe, the tattoos multiply' (Bronnikov 1993: 50). In minimum and medium security prisons, the majority of inmates have tattoos (respectively 65 to 70 per cent and 80 per cent). Although it seems that female inmates are generally less likely to have tattoos, the percentage of tattooed female inmates is the highest in institutions with maximum security. Usually inmates create the tattoos themselves. The methods used are painful and primitive:

A single small figure . . . can be created in four to six hours uninterrupted work. The instrument of choice is a reconstructed electric shaver to which prisoners add needles and an ampule with liquid dye. Scorched rubber that has been mixed with urine is used for dye.

(Bronnikov 1993: 50)

It requires no further explanation to understand that the medical risks are significant. Bronnikov observed that white-collar delinquents and those who were convicted for political reasons do not sport tattoos. The rest of the inmate population hold these inmates without tattoos in contempt. When newcomers enter the institution they look at the tattooed inmates with respect and fear, because they understand that these inmates are experienced and will therefore know the drill. In general, it can be stated that the position of a prisoner inside the institution depends, among other things, on his experience, professionalism and knowledge of committing (certain) crimes and life in prison. Those who are 'able' to read tattoos can come to an understanding of the ranks and divisions among inmates. It is important to come to an understanding of symbols and their places on the body. For example, an eight-pointed star on the chest means that the inmate is a professional delinquent. However, if he sports an eight-pointed star on his kneecaps, the inmate reveals that no-one can reduce him to his knees. According to Bronnikov, finger tattoos are the most common, whereby each finger tattoo stands for a particular conviction. Pictures and symbols represent the crimes committed. Wearing these symbols wrongly can lead to the killing of the inmate by his fellow inmates. The finger tattoos are always visible (tattoos on other body parts are usually covered by clothing). The same is true for facial tattoos that are often the sad result of the 'card game of chance'. If a loser is not able to pay his debt in money or in other valuables, the winner can decide to make him his slave, sexually abuse him or force him to get his face tattooed (with a swastika, prison bars or, for example, the word 'slave'). In the prison hospital, these tattoos can be removed and after the operation the loser can be transferred to another institution and escape from his creditor(s). Inmates at the bottom of the social ladder, usually delinquents who have been imprisoned for sexual offences, are often tattooed against their will.

Tattoos in American prisons

It is not entirely clear when American inmates started tattooing themselves for the first time. According to DeMello (2000) there are great similarities between the emergence of prison tattooing and biker and Chicano tattooing. It seems that biker tattoos often have an anti-social character (e.g. 'born to lose' and 'fuck the world'). The tattoos are placed on visible parts of the body. They function as a way into the biker culture, as a private expression and as a public comment. Chicano tattooing emerged in the 1940s and 1950s and must be placed within the context of Hispanic gangs in California, Texas, New Mexico and Arizona. Although nowadays colour is also used by Chicano and Mexican tattoo artists, classic examples of Chicano tattoos are – just as biker tattoos tend to be – exclusively done in black ink. Images often used are Christ, the Virgin of Guadeloupe and the name of the hometown or neighbourhood. These tattoos function as a way of identifying gang members and often express loyalty to the community, family, women or God (DeMello 2000). The following case deals with the tattoos of inmates that are members of 'Nuestra Familia', a California-based prison gang.

Gang tattoos in an American prison

Michael P. Phelan collected data in the California State Prison while working there as a correctional officer (1984–90). The sketches he collected were related to different gangs – one of them being the Nuestra Familia. It is believed that this gang was founded in San Quentin prison, but today members are recruited amongst Mexican Americans from the northern part of California. Nuestra Familia has developed into a highly structured organization with military traits. In charge is a General with unlimited and absolute power. Then there are ten Captains: the first Captain is the successor to the General. The Captains supervize Lieutenants and soldiers. In order to be eligible for promotion to Lieutenant, a soldier must have killed three people and to achieve the position of Captain this number goes up to five. A current member must sponsor individuals, who are willing to join this gang. Membership is considered to be a lifelong commitment.

According to Phelan et al. (1998) five 'career stages' can be distinguished in the 'curriculum vitae' of a Nuestra Familia member: pre-initiate, initiate, member, veteran and superior. During the first phase, pre-initiate, an individual can only claim rudimentary group affiliation. By wearing simple tattoos – e.g. the Spanish word Norte (North) that refers to the territory of the gang – an individual can demonstrate interest in becoming a member. During the initiate stage, more commitment to the gang is required. A typical tattoo worn by initiates is a rose: '. . . the rose without more advanced Nuestra Familia markings, suggest to correctional officers that the individual wearing such a tattoo might be actively seeking to prove himself. This leads "wise" correctional officers to approach such an inmate with increased caution'(Phelan et al. 1998: 286).

Being an initiate, the individual has to seek or create opportunities to show off his commitment to the gang. The initiation ends when membership has been approved of. A member can wear the letters NF indicating Nuestra Familia membership. Membership tattoos do not give information on the owner's 'curriculum vitae'. Veteran tattoos shed a light on personal accomplishments. There are, for example, tattoos that depict how often the wearer has been imprisoned or has killed somebody (i.e. the number of teardrops always worn under the left eye). Veterans also wear simple stars. A star on the arm or the body indicates that the wearer has killed one person, a star worn on the face means that the owner has killed twice. Veteran tattoos do not generally offer insight into the rank of the wearer in the gang. Some tattoos, however, are reserved solely for the elite. Only Lieutenants, Captains and the General are allowed to wear a tattoo of 'a Mexican male with a large moustache, rifle, bandolier and a sombrero that covers his face and hides his identity. The number of bullets tattooed on the bandolier represent how many kills the person has to his credit' (Phelan *et al.* 1998: 291).

Conclusions

The examples described above demonstrate – just as Sanders (1989) observed – that an important function of tattoos worn by inmates is found in the construction of a personal identity that resists the homogenizing pressure of the prison system. The ultimate way of demonstrating this control over and sovereignty of the own body is by personally modifying it through the painful application of tattoos. Those who have been tattooed against their will – in the way described by Bronnikov in the Russian case – are not in control. They are considered to be 'weak' and 'unmanly'.

Deciding to inscribe the body with a tattoo can be an individual decision, which is based on coming to terms with the institutional attack on one's personal identity. Another way of dealing with this kind of strain is the formation of a subculture among inmates, like a prison gang, that tries to resist the pressures of the institution. Both cases provide examples of counter-cultures. In these counter-cultures, such as the Nuestra Familia in the US, the tattoo reveals hierarchical relations and thus functions like a uniform that gives us information about accomplishments and rank. In this collective, men can feel strong and, in the presence of outsiders, they act as one. They can show off strength and dominance in an aggressive manner.

Mathiesen (1965) noted that inmates experience the space of prison relative to their personal experiences, which have occurred in space outside or inside the institution. The ways in which inmates try to fight homogenizing pressures on an individual and collective level, are related to their focus on experiences in the space of prison. When inmates compare their present situation with developments outside the space of prison, they tend to focus on their marginalized position in society. The tattoo is traditionally more often present in the lower social strata of society. These 'imported' tattoos also help the inmate to protect his personal identity. They confront the wearer with his 'roots', e.g. being a biker or a Latino in space outside prison. No board of prison directors can take that away from them.

Perhaps in the Netherlands inmates do not need tattoos to fight the pains of imprisonment and the stripping of their personal identity in the way described above. In addition, the pressure to conform to the demands of prison subcultures may be lower as they are less pertinent in the Netherlands (see, for example, Grapendaal 1987). However, further research will be needed to verify this. In addition, research needs to be conducted into alternative ways of identity formation and resistance in prisons in the Netherlands.

Notes

1 In real life, however, most inmates would rarely do so.
2 In the Netherlands, for example, about 5 per cent of the total prison capacity is destined for women (Janssen 1994; 2000).
3 'Tattoo' is derived from the Polynesian word 'ta-tu' or 'tatau'. It means among other things beating or scratching (Schuler 1997; Sanders 1989).
4 Christian pilgrims, for example, received tattoos as souvenirs of their pilgrimages to the Holy Land. The Celts were familiar with the art of tattooing, even before the Roman conquest (MacQuarrie 2000).
5 See also: Garland Thomson, 1996.
6 See, for example, Durkin and Hougton on stereotypes formed by children and adolescents of people with tattoos as being delinquent (2000). Further to this, see Bekhor *et al.* on employer's (usually negative) attitudes toward people with visible tattoos (1995).
7 In the early nineties of the last century I looked into the way female inmates from Latin America experienced being detained in the Dutch prison system (Janssen 1994). In 2000 I finished a research project on the effects of short-term imprisonment of male delinquents in the Netherlands (Janssen 2000).
8 I would like to thank Simone van der Zee for helping me to gain access to this information.
9 In prison records it is only mentioned if an inmate has tattoos or not. On the registration form is a category 'skin'. The administration can enter 'tattoo(s)' as a distinguishing feature of the skin.
10 Aside from Bronnikov another Russian official, Sergejewitsch Baldajev, 'collected' hundreds of inmate tattoos during the Soviet era (Schuler 1997).

References

Bekhor, P.S., Bekhor, L. and Gandrabur, M. (1995) 'Employer attitudes toward persons with visible tattoos', *Australian Journal of Dermatology* 36: 75–7.
Bhavnani, K.K. and Davis, A. (1995) 'Incarcerated women: transformative strategies', *Gevangen vrouwen. Over criminaliteit en detentie*, Amsterdam: Nemesis.
Bronnikov, A.G. (1993) 'Telltale tattoos in Russian prisons', *Natural History* 11: 50–8.
Caplan, J. (2000) 'National tattooing: traditions of tattooing in Nineteenth-century Europe', in Caplan, J. (ed.) *Written on the Body: The Tattoo in European and American History*, London: Reaktion Books.
DeMello, M. (2000) *Bodies of Inscription: A Cultural History of the Modern Tattoo Community*, London: Duke University Press.
Duncan, N. (ed.) (1996) *Body Space: Destabilizing Geographies of Gender and Sexuality*, London: Routledge.
Durkin, K. and Houghton, S. (2000) 'Children's and adolescents' stereotypes of tattooed people as delinquent', *Legal and Criminological Psychology* 5: 153–64.
Garland Thomson, R. (ed.) (1996) *Freakery: Cultural Spectacles of the Extraordinary Body*, New York: New York University Press.
Goffman, E. (1977) *Totale Instituties*, Rotterdam: Universitaire Pers Rotterdam.
Grapendaal, M. (1987) *In Dynamisch Evenwicht: Een verkennend onderzoek naar de gedetineerden-cultuur in drie Nederlandse gevangenissen*, WODC 78. Den Haag: Ministerie van Justitie.

Grosz, E.A. (1995) *Space, Time and Perversion: Essays on the Politics of Bodies*, New York: Routledge.

Janssen, J. (1994*) Latijnsamerikaanse drugskoeriersters in detentie: ezels of zondebokken?*, Arnhem: Gouda Quint BV.

Janssen, J. (2000) *Laat Maar Zitten: Een studie naar de exploratieve werking van de korte vrijheidsstraf*, Den Haag: Boom Juridische Uitgevers.

Klees-Wambach, M.-L. (1976) *Kriminologische und kriminalistische Aspekte des Tätowierens bei Rechtsbrechern*, Freiburg im Breisgrau: Albert-Ludwig-Universität.

McDowell, L. (1996) 'Spatializing feminism. Geographic perspectives', in Duncan, N. (ed.) *Body Space: Destabilizing Geographies of Gender and Sexuality*, London: Routledge.

MacQuarrie, Ch.W. (2000) 'Insular Celtic tattooing: history, myth and metaphor', in Caplan, J. (ed.) *Written on the Body: The Tattoo in European and American History*, London: Reaktion Books.

Mascia-Lees, F.E. and Sharpe, P. (1992) 'The marked and the un (re) marked: tattoo and gender in theory and narrative', in Mascia-Lees, F.E. and Sharpe, P. (eds) *Tattoo, Torture, Mutilation and Endorement: The Naturalization of the Body in Culture and Text*, Albany: The State University Press of New York.

Mathiesen, T. (1972) *The Defences of the Weak: A Sociological Study of a Norwegian Correctional Institution*, London: Tavistock Publications.

Morgan, D. (1993) 'You too can have a body like mine: reflections on the male body and masculinities', in Scott, S. and Morgan, D. (eds) *Body Matters: Essays on the Sociology of the Body*, London: The Falmer Press.

Phelan, M.P. and Hunt, S.A. (1998) 'Prison gang members' tattoos as identity work: the visual communication of moral careers', *Symbolic Interaction* 21(3): 277–98.

Reiman, J.H. (1995) *The Rich Get Richer and the Poor Get Prison: Ideology, Class and Criminal Justice*, Boston: Allyn & Bacon.

Rose, G., Gregson, N., Foord, J., Bowlby, S., Dywer, C., Holloway, S., Laurie, N., Maddrell, A. and Skelton, T. (1997) 'Introduction', in Women and Geography Study Group, *Feminist Geographies: Explorations in Diversity and Difference*, Harlow: Longman.

Sanders, C.R. (1989) *Customizing the Body. The Art and Culture of Tattooing*, Philadelphia: Temple University Press.

Schrader, A.M. (2000) 'Branding the other/tattooing the self: bodily inscription among convicts in Russia and the Soviet Union', in Caplan, J. (ed.) *Written on the Body: The Tattoo in European and American History*, London: Reaktion Books.

Schuler, D. (1997), 'Ta'tatau, ta'tatau . . . De oorsprongen van tatoeage in Europa', *P & M* 78: 17–30.

Steward, S. (1990) *Bad Boys and Tough Tattoos: A Social History of the Tattoo with Gangs, Sailors and Street-Corner Punks, 1950–1965*. New York: Harrington Park Press.

Thomas, C. (1977) 'Theoretical perspectives on prisonization: a comparison of the importation and deprivation models', *The Journal of Criminal Law & Criminology* 68(1): 133–45.

Thomas, C., Petersen, D. and Zingraff, R. (1978) 'Structural and social psychological correlates of prisonization', *Criminology* 16(3): 383–93.

Vail, D.A. (1999) 'Tattoos are like potato chips . . . you can't have just one: the process of becoming and being a collector', *Deviant Behavior: An Interdisciplinary Journal* 20: 253–73.

Further reading

Dutton, K.R. (1995) *The Perfectible Body: The Western Ideal of Male Physical Development*, New York: Continuum. This fascinating book looks into (the history of) ideas related to the state of male bodily perfection. Among other things, the author explores the world of body building. He explains how the muscular body has been presented in the world of high art and that the developed body has become part of popular culture.

Miller, S.L. (ed.) (1998) *Crime Control and Women: Feminist Implications of Criminal Justice Policy*, Thousand Oaks: Sage Publications. In my chapter, I explained that research in penology often focuses on men. In this book the authors look into the effects of criminal policies in the US in the 1990s on women and children.

15

'YOU QUESTIONING MY MANHOOD, BOY?'

Using work to play with gender roles in a regime of male skilled-labour

Rhys Evans

Summary

This chapter examines the way in which regimes of work and discourses of masculinity are entwined, whilst showing how, by appealing to contradictory aspects of different discourses, individuals are empowered to engage in behaviours which violate the strict limits of either. It discusses and demonstrates an auto-ethnographic method.

- The auto-ethnographic method
- Staple economy, skilled labour, and masculinity
- Stereotypes of rural masculinities
- Performativity.

Introduction

The statement in the title was chosen because it exemplifies the two main points I wish to make in this chapter: that there are a multiplicity of discourses of masculinity which can be seen as in competition with each other and which, narrowly defined, offer men a kind of agency which enables them to engage in 'un-manly' activities without diminishing their manliness. The second is that critical self-examination through the use of an auto-ethnographical method can be a useful tool in examining and representing social constructions of masculinity.

This chapter considers discourses of masculinities in the rural *staples* economies of Canada, exploring how the cultures of a particular productive regime produce specific masculinities. I say *rural* staples economies because traditional rural occupations – farming, logging, fishing – tend to be in the staples sector. These occupations share characteristics which, I will demonstrate, form a *culture* and *regime* of masculinity. Of course, these jobs are not exclusive to rural areas – but they do form the majority of what have been traditionally regarded as rural livelihoods. I then briefly examine *performativities* of gender based upon Judith Butler's definitions (1993) in order to demonstrate how these masculinities become seen to be natural. I use an auto-ethnographic method based upon my own experiences during 20 years as a truck driver. And finally, I try to identify men as gendered persons, rather than a homo-generic category of human beings, asserting that we need to remember that the lives of working men deserve the same careful and critical enquiry we now assume should be given to working women. I do so in order to unpack stereotypes of working men, to both understand and empower the individual, and to aid our understanding of the structuring of, in particular, rural masculinities, so that we might better understand its impact on rural organization and change.[1]

The auto-ethnographic method

I must first address my use of *auto-ethnography* in this chapter. I take the term from Mark Neumann (1996). He claims that the critical intersection of works of *auto*-biography and *ethnography* offer a useful way into studying the relationship between individuals and the cultures within which they are embedded. By situating the author within the texts, the subjective, contextual nature of the ethnographic project is highlighted and the author's own assumptions can be openly interrogated. There is an established tradition of advocating critical self-reflection within feminist litera-tures (Pratt 1986; Rose 1993; Hughes 1997) and within certain communities of ethnography (Ellis and Bochner 1996; Denzin and Lincoln 1994); the present project attempts a similar approach to understanding.

Even with the example of such literatures, however, it is obvious that in using an auto-ethnographic method I run the risk of 'being seen by colleagues as [an] emo-tional exhibitionist' (Ellis and Flattery 1992). Following Pratt (1986), I wish to study gendered subjectivity and, as 'subjectivity is situated such that the voices in our heads and the feelings in our bodies are linked to political, cultural, and histor-ical context' (Ellis and Flattery 1992: 4), critical self-reflection upon my own experiences of working-class masculinity offer an appropriate means of entry into the topic.

My truck-driving stories are also appropriate at another level. This is because story-telling is an important facet of the trucking culture that I grew up in. Due to the isolation and loneliness of that life, story-telling plays a central role in intersub-jective contact in truck stop cafés or during the other brief periods drivers have in the cab of their truck. Because time spent in the cafés is short, those who go on for too long will often be interrupted and someone else will take over the story-telling. So stories become revisited, refined, rubbed and polished by the driver when alone at the wheel of his truck. Telling oneself the stories is an antidote to the boredom which accompanies the job, and allows the driver to perfect them until they exist in an almost essential form. Just as a field researcher reads and re-reads his or her field notes to refine them into the stories they wish to tell, truckers will work their stories until they can express the largest possible amount with the minimum use of words.

Refined this way, the stories tend to fall into a number of almost archetypal categories. These include themes such as *my brush with death and how I survived; my brush with authority and how I 'fixed his little red wagon'*; or *my brush with sex and how I acted like a real man*. I too became a good storyteller and that is one of the skills I bring to academic work. Now as I return to my trucking stories I can see how my participation in the process and the stories themselves were structured in certain ways – ways of masculinity among others. My 'field notes' are written in my memory, having been revisited dozens and dozens of times. I propose to look at them again, this time from the perspective of academic theories of subjectivity.

Staples economy, skilled labour, and masculinity

I take the term staples economy from Harold Innes' use of it in *The Bias of Communication* (Innes 1951). A Canadian economic historian, he used the term to describe economies where the primary economic activity involved the production of raw or semi-processed commodities in distant rural areas. These staple products could be agricultural, mineral or forest products. His work established the interpenetration and mutual dependence of the urban core and rural hinterland, focusing on the exchange of goods, wealth and ideas between the two.[2]

Although staples economies differ spatially and temporally, they share a number of features, including isolation from urban centres, working within an often difficult and sometimes dangerous physical environment in order to produce commodities from it, and ruggedly masculine occupations such as logger, miner, farmer, truck driver, rail worker, roughneck, etc.[3] Staples economies tend to produce much wealth because of the productivity of the enterprises: they involve massive quantities of commodities, whether wheat, timber or ore, but the work is mechanized and physical labour is replaced with machine inputs in order to achieve economies of scale. From the perspective of my own experience in Western Canada, these jobs tend to involve isolation, long hours or monotonous work, working with machines at the limit of their technological capacities, as well as physical danger from the natural environment and the industrial processes themselves. Because of the large capital inputs necessary to organize these endeavours, the jobs tend to be unionized and repetitive and pay good wages due to the high productivity and importance of the commodities to urban life. The routinization of the work means that much of it is dependent upon skilled labour, combining hard physical work with intimate knowledges of the work process.

In a staples economy, male work tends to be working class. But it is a *skilled* working class and the result is a regime of what I call *skilled labour*. Whether farming, logging or mining, productivity is high and the rewards are thus relatively substantial. By specializing in certain mechanical aspects of the production process and by being a member of a unionized work force, men working within a staples economy can earn considerable incomes, a factor which increases their self-respect.

My interest is in an ethos of masculinity which pervades rural cultures in the developed world. These constitute a regime of masculinity built from the historical, economic and cultural factors present in a place. There are a number of qualities of masculine character which devolve from the way in which these working men go about their business. Distance and isolation drive one aspect of work-based masculinities – the need for self-reliance and to possess a diverse set of skills and behaviours in order to get the job done. These involve the ability to keep the equipment working, which enables such high productivity. It also reflects a general lack of access to specialists and the need to perform other tasks which are mundane and which involve *not-so-skilled labour*, tasks which cannot be contracted out for reasons of cost or availability of external labour. All that matters is that the commodities are produced, that the goods are delivered to the customer. This is the touchstone, the ultimate guarantee of a man's employability and earnings, and ultimately his masculinity – can he produce the goods? And in order to 'produce the goods', a man does whatever he has to do in order to get the job done. This requires a much greater range of behaviours than usually ascribed to working men and this also leaks into the private and domestic sphere, encompassing behaviours which do not fit the stereotype of 'manliness' yet which are necessary in order to 'produce the goods' in both commercial and private life.

I have run up against the limits of technological specialization in the productive process many times. Working as tank truck drivers for a major petroleum retailer, we used extremely expensive specialized trucks and trailers to deliver 52,000 litre loads to individual stations. This was accomplished by pulling two 32-foot trailers which were loaded in state-of-the-art electronically controlled bulk plants. Delivery was accomplished by the use of gravity feed from belly-mounted spigots featuring a double valve system for safety. In the post-Exxon Valdes era, not spilling product was sacrosanct – any spill over 50 litres had to be reported to Environment Canada. As this involved much paperwork and incident inspections, it was vigorously discouraged by management.

The trouble was that occasionally you would get a delivery where, the total amount in the truck tank had to be delivered into two ground tanks as neither had the capacity to take the whole load. At that point, all the high-tech systems broke down. You would have to manually hold a knife valve half closed and watch for a sign that the tank was full in the small sight glass in the spigot. When that happened, you had about one and a half seconds to slam the valve shut before gasoline began spilling out over the ground. In this way, one quarter of a million dollars of high-tech equipment was reduced to being controlled by a guy with a wrist watch and an anxious eye.

In staples economies, many of the jobs are physically dangerous[4] and the dangerousness increases as skill levels decrease. Part of the skills acquired in skilled labour work consist of knowing how to avoid excessive physical work,[5] either lifting heavy objects or having to manoeuvre them into position. By doing things in the right order, knowing what the problems might be and planning ahead to avoid them, the daily grind – the sum of physical and mental labour of the day – could be moderated. As a young truck driver, for example, I soon learned that I had to know the location of every address which had a delivery, as well as where the delivery was on my truck. Otherwise I would be looping back, hurrying up to get finished on time, or 'humping' freight back and forth from the front to the back of the truck in order to get at a certain delivery at a certain time. Once tired and in a hurry, it would be very easy to hurt my back, have a traffic accident or otherwise be injured. Likewise, forcing a tool puts the worker at risk of harm as he or she is always softer than either the tool or the object to which it is applied.

I worked for a railway company as a truck driver, up in the north of British Columbia. In quiet times I would watch the Section Gang crew repairing the tracks. One day I was watching a young guy struggling manfully to hammer spikes into particularly tough ground. It was summer and it was very hot. He was soaked with sweat and clearly tired, hammering away with a ten pound mallet at the recalcitrant spikes. Each spike would take many blows to get into the ground and, as he became more tired, his aim became less accurate. Finally one blow missed and the mallet bounced up off the steel rail and onto his leg. He screamed with pain and set about displaying his newly acquired collection of expletives.

An old veteran ambled over to him and, looking at him, shook his head. 'You're not holding your mouth straight' he said with a wry grin. The old guy picked up the mallet and set a spike. With three easy, long swings the spike disappeared into the plate. The young guy and I couldn't believe our eyes. The old fellow looked up and said, 'It's easy, you just let gravity do the work, all you do is tell it where to fall', and walked away. Through the practice of specific skills, he had 'mastered' the job and could display his mastery through not breaking into a sweat, or risking injuring himself with the hammer.

This acquisition of skill came through the performance of the job, through reflection on what was encountered and through repeated practice. The motivation for learning this came from many sources – the personal wish to have a day go 'easy'; peer pressure to look cool and calm. Because the display of these skills – of getting the work done in an unruffled way – gained respect, it was a positive indication of masculinity. Due to the reticence of our colleagues, the rewards for this skill acquisition were usually only indirectly external.[6] Rather, the quiet satisfaction felt when accomplishing a particularly difficult task with efficiency and smoothness, the 'pride in the job', were the main rewards for acquiring the skill. Verbal acknowledgement from others was rare, which increased its value, but each task accomplished skillfully, at least initially, could be a reward in itself.

For example, on Vancouver Island I had a job hauling bulk timber from the mills on the Island to tidewater ports. We would haul 45-foot semi-trailer flat decks with lifts of rough-milled dimensional lumber. These loads would be about 11 to 12 feet high. They had to be secured to the trailer with at least two chains each, in order to assure they remained where they were put. This meant employing a minimum of six chains. To lose a load would cost a driver a large fine, cause trouble with the company and risk killing someone else on the road. The chains were heavy logging chains and one was required to throw them over the load, fasten them on the other side and fasten them tight with a steel cinch. The chains weighed about

200 lb in their entirety as they were about 50 feet long. To get the chains over the load you had to employ a trick — a trick no one told me about. Not every truck used the chains and not every load required them either, so this wasn't something I had encountered before.

For me it was extremely frustrating. I would gather up the chain and manfully heave it over the top of the load. Only about half of my attempts were successful, however. The rest of the time, the chain would tangle into a big ball and come streaking back down my side of the trailer, barely missing my hard-hatted head and I would have to pick it up again, loop it up and heave yet again.

Finally, weeks after I started, a small Sikh guy took pity on me and showed me that there is a right way and a wrong way to coil the chain and that, if you do it the right way, even a little guy like him could get it over the load every time. It took another week to get it just right — 'just right' meaning that I could do it without sweating or getting dirty — but from then on, it was one of the repertoire of skills which were available to me. No longer did I hold up the queue of trucks and no longer did I feel a 'greenhorn'. By performing this (and other) essential skills I became 'one of the guys'. Although there was no verbal reinforcement, I nevertheless knew when I had been accepted as a truck driver, and as a man. Men displayed a lack of sweat and a lack of concern as they went about their business. Once accepted, one was offered the respect of (ironically) small intimacies, small acknowledgements and welcome into the serial conversations which formed social interactions in a job which was always on the move.

For me, one of the main attractions of the job was the lack of surveillance it entailed. Once a driver left the company's yard in the morning, the bosses did not see him or her again until they came in at the end of the shift. What mattered wasn't what you said, or even, to a large extent, what you looked like (I never had a problem with long hair and a beard, for example, even though it was unusual in the early 1970s). What mattered was that you delivered the goods – that your work was timely, efficient, that the customers had no reason to complain, and that you treated the equipment well. Those were the defining qualities of skill. The union protected you from discrimination over looks or political opinions but it wouldn't protect you from being unskillful. As long as you got the job done, you were considered, in this very masculine occupation, to be a man.

This was a realm of skilled unionized labour so that, although I spent my work hours as a truck driver, once I left the job, I was my 'own man'. In a very Thoreauian sense (1960), once my labour was done, the company had no purchase on the rest of my life. I 'worked to live', not 'lived to work'. The work supported the rest of my life. Not that I did not consider myself a truck driver in the rest of my life, but driving trucks was what I did for a living, not the sum total of who I was. And on Vancouver Island, I found I had much in common with the drivers, loggers and fishermen whom I met. We worked to make a living, but we also built our independence around an extensive life outside of our occupation and that was as much a part of who we were – and our sense of ourselves as men – as our work life, whilst at the same time also being a factor of our occupational masculinities.

Of course there were degrees of being a man. In trucking these were not so much organized around wearing cowboy boots or speaking on the CB-radio (in fact, I seldom had a CB-radio in my jobs). Rather they were organized around one's ability to work long hours (endurance), to perform the necessary tasks skillfully (to guide the truck into difficult places with no accidents or damaged loads), and to respond with a certain degree of humility and humour to the difficult situations one faced. This last was very important – the demeanour which indicated one's 'toughness' and masculinity was not machismo, but rather a self-derogatory sense of resilience and independence which proclaimed that no matter how hard the work was, you could keep your distance from it and still keep a sense of humour. I regularly worked 70- to 90-hour weeks, a condition forced by the distances involved in Canada and the United States. What was important was to not complain about it too much, and if one did, to do it with good humour.

Stereotypes of rural masculinities

Stereotypes of rural men tend to portray them as possessing a super-masculinity, a form of *machismo* which is built of hard physical labour in difficult or uncomfortable circumstances. This, especially when combined with the traditional gendered division of labour, limits the accepted range of masculine behaviours. Thus, rural men are often seen to be more overtly masculine than their urban counterparts and even less likely to get involved in the domestic sphere. The gendered division of labour would appear to be more extreme than in urban situations.[7]

Yet these stereotypes are seldom as absolute as portrayed. In particular, the range of things a man will have to do in order to survive is much greater than the stereotypes allow. Many rural men, whether farmers or loggers, engage not only in plenty of 'scut work', that is, demeaning labour, but also in acts of tenderness and nurturing; many have a keenly developed aesthetic sense in terms of the landscape or of domestic and wild animals.[8] And many play a more active part in the raising of their children than is commonly acknowledged. This is partly due to the remote nature of rural spaces where family recreation often takes place within the immediate environment. For example, the same areas where logging is conducted also contain lakes, streams and other features which lie at the heart of family recreation.

Also, children can have much more access to their fathers within a rural staples economy. The separation between work and home can be transgressed in environments where surveillance is difficult or where the productive spaces belong to the worker, such as on farms, in the family business, or in the immediate surrounding environment of a small logging outfit. Children (boys *and* girls) are trained in the skills of farming by their parents; truckers take their children with them to work in their trucks, especially during school breaks when the mothers may be working outside the home;[9] children play around and ultimately help out in family enterprises like repair businesses, bed and breakfast accommodation or any other enterprise situated on the family landholding.

When the fathers are involved in these domestic tasks, what legitimates these acts as acceptable male behaviours is the emphasis on 'producing the goods',[10] whether keeping the farm in the family or keeping the truck running during the summer. If masculinity is quintessentially measured by this yardstick, it justifies a number of behaviours that stray from strict notions of what is usually considered masculine behaviour.

To recap, I want to highlight several features of the discourses of masculinity which I encountered when immersed in the work of a rural staples economy. The key features, as I see them, of 'being a man' were *skill, endurance (toughness)*, and *independence*. Men who displayed these characteristics were respected. That respect was shown not by verbal accolades, but by acceptance into a community of others and by the absence of dismissive ridicule.[11] The ultimate accolade, seldom given, was 'he's a real man', often delivered in a depreciating, humorous sense.

Interestingly, what possession of the above qualities made unimportant were appearance, origin, literacy, or politics. You could have long or short hair, be white or Sikh, be illiterate or attending a university, be right wing or left wing. As long as you displayed your masculinity through a display of the above characteristics (skill, endurance and independence), the other features remained incidental. In other words, as long as you could get the job done, it didn't matter what you looked like or what you believed in. This then is an alternative discourse of masculinity, one that is constituted partly from norms of comportment, but also as a result of being located in a particular place and culture, one that grew out of generations of skilled labour within a particular regime of production.

Performativity

The emphasis upon the relationship between performing work and masculine identity leads us to consider the often inaccurate use of the term 'performativity' in discourses of identity. The term has a specific and fairly narrow definition and I wish to distinguish between more general notions of performance, and performativity in the sense of social actors enacting the possibilities provided by discourses of, in this case, masculinity. One 'performs' work, but the performativity of masculinities is a process out of which identities are constructed; and it operates both to legitimate and to conceal the legitimization of certain behaviours and traits. Thus, discourses of masculinity appear natural and rooted in history, having the force of givens, not to be questioned. Interestingly, this process of legitimization and dissimulation also operates to prevent men from achieving ultimate masculinity and what keeps them striving toward it at the same time.

Much of the understanding of gender identity to have come out of feminist studies is based upon psychoanalysis, referring back to Freud and passing through the works of Lacan (1978, 1985) (see also Butler 1990, 1993; Kristeva 1980; Fuss 1993). This is particularly true of *subject-formation*, i.e. the creation of the individual subject, the part identified as the *self*. I do not wish to repeat a long psychoanalytic exegesis here. It is, however, germane to the issue of performativity, since the Lacanian subject is never perfect and unitary, but rather is engaged in a constant struggle with *lack*. For Butler this means that 'heterosexual performativity is beset by an anxiety that it can never fully overcome, that its effort to become its own idealization can never be finally or fully achieved' (1993: 125). This lack opens an entry point into the myth of masculine homogeneity; it allows us to see men as constituted through their identifications and actions in social space rather than as simple stereotypes of masculinity.

Masculinity is a performed social identity rather than a state of being. The routine performance of their lives means that, 'most men's lives reveal some departure from what the "male sex role" is supposed to prescribe' (Carrigan *et al.* 1987). It is in this performance that the performativities of masculine gender can be subverted. Also, being born with a penis is not enough. There are hierarchies of masculine performance and exceptional performance is equated with exceptional masculinity. How well one performs tasks and the actual performance of those tasks determines one's place in a masculine hierarchy, and that place is never full or permanent. For men, as well as for women, in other words, *not only must gender be done, but it must be seen to be done, again and again* – it must be iterated and, performed within a social space.

Discourses of masculinity and its performativities occupy the role allocated to 'male gaze' in Iris Young's (1990) schema.[12] The male gaze is turned upon other men (and ourselves) in terms of 'manliness'. It is thus difficult to see how men can ever possess a static unitary subject position; rather they are always striving through uncertainty to achieve it.

The clear route to achieving masculinity is never quite within reach, it remains knowable only in part. Only through repeated iterations of male performativities can a man feel comfortable or settled in his masculinity. Masculinity can only be 'stored' for a very short while, and masculine subjectivity must be constantly enacted; a fall from grace is always possible if the performance suffers.

Discourses of self-identity only have the power to affect individuals and collectivities when they are expressed as acts in the real world. Yet acts themselves only acquire meaning through discourse. Discourse operates at both the level of the collective, and of the individual. Social judgements upon a man's masculinity ultimately reinforce his own perception of his performance. It is not even he himself who provides the judgements of his own adequacy, but rather the dissimulative nature of performativity itself. One of the virtues of a discourse of masculinity built around concrete job performance is that, once a number of 'jobs' have been performed, the actor can take comfort in his performance as a defence against the ambiguous and contradictory nature of the discourses.

The *hierarchy* of masculinity also has its genesis in the nature of performativity itself. Because of the temporal and contingent nature of it, there is a slippage 'between discursive command and its appropriated effect' (Butler 1993: 122). That slippage means that all men can be, to a greater or lesser extent, manly, depending upon how their performances match the 'discursive command' or ideal. And within the discourses of masculinity, some social acts accrue more value than others. Thus an actor can choose to act unmanly, knowing that the particular act carries less weight than others. A man can make an assertion of masculinity in a certain way to defend a particularly unmanly style of comportment. This allows men to engage in unmanly behaviours without serious threat to their manhood through the citation of the performance of other, more positive aspects of the discourses of masculinity.

For Butler, 'such acts, gestures, enactments, generally construed, are performative in the sense that the essence of identity that they otherwise purport to express becomes a *fabrication* manufactured and sustained through corporeal signs and other discursive means' (1990: 336). Yet, as we have seen, there is more than one way to *act* like a man. Work tasks, like throwing chains over loads, shifting gears without grinding them, and most importantly, coming back at the end of a shift with all the work done, represent another type of performing like a man which does not involve the direct *fabrication* of the self, or if it does, it does so indirectly. Thus, individual men can use one aspect of performing masculinity to subvert the demands of another. Which aspects of the performance of masculinity you choose to emphasize makes a difference to what kind of man you are; and it is this which allows those of us who do not look and act like the 'Marlborough Man', for example, to consider ourselves, and be considered, as men.

To an extent then, different aspects of performing masculinity contradict one another. There is little room for the John Wayne type hero clearing out a jam which has shut down a big machine. The work is dirty and no one notices what you do, except when the machine stops. The qualities needed to keep a machine producing are not necessarily those of the 'hero'. When masculinity is constructed around set definitions of performance, such as getting the job done, this can be used against other criteria. In a similar way, discourses of masculinity can also contradict each other. Whilst no actor is entirely freed from the dissimulative nature of performativity, there are times and occasions when an actor can choose between discourses in order to transgress the dictates of one or the other. By emphasizing the prime factors in the discourse, transgression of secondary ones does not necessarily diminish a person's manliness. The relationship between the performance of tasks and the representation of the self can be used as a strategic resource by those who position themselves as other to the mainstream in order to support their entitlement to identities which include masculine occupational categories.

Conclusion

Rural working men tend to be seen as stereotypes or ciphers – macho, hard and relatively unsophisticated. The range of behaviours available to rural mainstream men is framed and bound by the discourses out of which they are constructed. Yet these discourses offer definitions of masculinity which are much more subtle and empowering than the standard stereotypes of a macho 'Marlborough Man'. Performing work offers rural men ways of incorporating behaviours, feelings and tendencies which confound the stereotypes through the use of work performance to occupy an unchallengeable position of masculinity which remains undiminished by tenderness, gentleness, confusion and insecurity – in other words, the whole range of human feelings and emotions. Like so many subjects of social science enquiry, working men confound our stereotypes and respond to the difficulties of wresting a living from the world with the full range of human emotions and aspirations. Specific discourses of masculinity have sedimented into cultures and regimes of masculinity in rural staples economies, discourses which may surprise with what they enable, but which are nevertheless key components of individual men's personal identity. A further question which requires more enquiry is whether these masculinities can survive the transition from a rural staples economy to the post-production, rural service economy which seems to be ascendant.

Notes

1 This transition from primary sector to service sector activity in rural areas forms the basis of a subsequent research project called 'Hell no! I'm not selling ice cream cones to a bunch of tourists – I'm a man, dammit!'.

2 This reflects the previous statement that this type of work is primarily rural, but not solely so.

3 I include truck driving within this category because, although it serves the manufacturing and urban-centre-distribution sectors of the economy, it also serves the primary production sector. In addition, the basic conditions of work in trucking resemble the machine-paced, socially-isolated work conditions of other skilled labour jobs. Even when working from urban locations, a truck driver escapes to the spaces between cities and thus is spatially differentiated from urban factory labour.

4 Not that sedentary middle-class jobs aren't dangerous, rather, in them the trauma tends to be temporally removed from the injury, e.g. heart disease, high blood-pressure, cancer, etc.

5 This isn't restricted to male work.

6 After all, individuals with greater and lesser levels of skill received the same rate per hour. Socially, the rewards consisted of negative feedback – the cessation of ridicule rather than overt verbal rewards.

7 Although this situation is changing in the urbanizing and technologizing countryside.

8 I know a number of retired loggers who have taken up landscape painting, for example.

9 Most women I know who drive trucks were taught by their fathers.

10 Especially if done within the environment of the family.

11 Interestingly, being accepted as 'one of the guys' often seemed to expose one to increased levels of teasing, but this was not dismissive. Rather it was a kind of training for the self-depreciating humour one was expected to display.

12 This raises the possibility that a 'female gaze' might also operate similarly.

References

Butler, J. (1990) *Gender Trouble: Feminism and the Subversion of Identity*, New York: Routledge.

Butler, J. (1993) *Bodies That Matter: On the Discursive Limits of 'Sex'*, New York: Routledge.

Carrigan, T., Connell, B. and Lee, J. (1987) 'Toward a new sociology of masculinity', in Brod, H. (ed.) *The Making of Masculinities: The New Men's Studies*, Boston: Allen & Unwin.

Denzin, N.D. and Lincoln, Y. (eds) (1994) *Handbook of Qualitative Research*, Thousand Oaks, Ca: Sage Publications.

Ellis, C. and Bochner, A.P. (eds) (1996) *Composing Ethnography: Alternative Forms of Qualitative Writing*, Walnut Creek, Ca: Sage Publications.

Ellis, C. and Flattery, M.G. (eds) (1992) *Investigating Subjectivity: Research on Lived Experience*, New York: Sage Publications.

Fuss, D. (1993) 'Freud's fallen women: identification, desire and a case of homosexuality in a woman', *The Yale Journal of Criticism* 6(1): 1–23.

Hughes, A. (1997) 'Rurality and "cultures of womanhood": domestic identities and moral order in village life', in Cloke, P. and Little, J. (eds) *Contested Countryside Cultures: Otherness, Marginalisation and Rurality*, London: Routledge.

Innes, H.A. (1951) *The Bias of Communication*, Toronto: Toronto University Press.

Kristeva, J. (1980) *Desire in Language*, New York: Columbia University Press.

Lacan, J. (1978) *The Four Fundamentals of Psychoanalysis*, New York: Norton.

Lacan, J. (1985) 'The meaning of the phallus', in Mitchel, J. (ed.) *Feminine Sexuality: Jacques Lacan and the Ecole Freudienne*, New York: Norton.

Neumann, M. (1996) 'Collecting ourselves at the end of the century', in Ellis, C. and Bochner, A.P. (eds) *Composing Ethnography: Alternative Forms of Qualitative Writing*, Walnut Creek, Ca: Sage Publications.

Pratt, M.L. (1986) 'Fieldwork in common places', in Clifford, J. and Marcus, G.E. (eds) *Writing Culture: The Poetics and Politics of Ethnography*, Berkeley: University of California Press.

Rose, G. (1993) *Feminism & Geography*, Minneapolis: University of Minnesota Press.

Thoreau, H.D. (1960) *Walden, or, Life in the Woods; and, On the Duty of Civil Disobedience*, New York: New American Library.

Young, I.M. (1990) *Throwing Like a Girl and Other Essays in Feminist Philosophy and Social Theory*, Bloomington: Indiana University Press.

Further reading

Jardine, A. and Smith, P. (eds) (1987) *Men in Feminism*, New York: Routledge. This is a good review of the numerous ways that men have approached feminist theory and practice. The volume encompasses a broad range of positions, some I personally find reprehensible. But it exposes the reader to the historical antecedents and early attempts to find a place for men in feminism, and highlights basic problems which still are relevant to men's and women's experience in coming to terms with patriarchy.

Luxton, M. (1980) *More Than a Labour of Love: Three Generations of Women's Work in the Home*, Toronto: Women's Educational Press. This piece of work from Canada shows the relationship between regimes of production (in this case, a mine in Manitoba) and the lives that are structured (constrained or enabled) by it. Its feminist approach to history and society is also clear and cogent.

Young, I.M. (1990) *Throwing Like a Girl and Other Essays in Feminist Philosophy and Social Theory*, Bloomington: Indiana University Press. Iris Marion Young's writing on the embodiment of gender discourse is one of the first pieces of feminist theory which spoke in a way that I as a man could see echoes of in my own life. *Throwing Like A Girl* is a good starting point to exploring the conjunction of theories of gender and embodied experience.

Part 5

SEXUALITY AND RELATIONSHIPS

16

EXPLORING NOTIONS OF MASCULINITY AND FATHERHOOD

When gay sons 'come out' to heterosexual fathers

Tracey Skelton and Gill Valentine

Summary

This chapter considers debates within masculinity and fatherhood studies with a particular emphasis on homophobia. Three in-depth case studies have been selected from a wider research project part of which focused on the marginalization experiences of young lesbian and gay people. The case studies have been selected for the insight they provide on 'coming out' processes within the family. The examples show how diverse heterosexual fathers' reactions can be when their sons come out to them as gay. It is clear that fatherhoods that are not constructed upon aggressively enforced heterosexual hegemonic masculinity are those that allow the space for a continuation, or even establishment, of positive and rewarding father–son relationships.

- Notions of masculinity/notions of fatherhood
- The research project
- Case studies of young gay men's coming out to heterosexual fathers.

Introduction

In this chapter we engage with some of the key debates within discourses on masculinity and fatherhood. Through an examination of selected case studies we consider a particular practice of power which is part and parcel of masculinity and fatherhood, i.e. the practice of homophobia.

Below we consider 'masculinity' and thence 'fatherhood'. In reality, of course, these two aspects of men's lives can, and for many do, overlap and intersect (Mac an Ghaill 1996b; Segal 1997). Our focus on homophobia as part of masculinity and part of fatherhood provides an important context for the father–son relations we examine in detail below.

The chapter interrogates the ways in which particular formations of masculinity and fatherhood can be examined through the lens of selected case studies located in the socio-spatial institution of the 'family'.[1] These case studies are of young gay men as they tell their 'coming out' stories and reflect upon the reactions of their parents, with a particular focus on the responses of their fathers. Hence we provide an ethnographic insight into a potential clash between two types of masculinity – heterosexual fathers and their gay sons.

Notions of masculinity/notions of fatherhood

Geography as a discipline is beginning to create the space for discussions about masculinity, male identity and men (Longhurst 2000). However, it is important that geography does not repeat the mistakes made in other subjects where masculinity has been presented as a 'monolithic, unproblematic entity' (Mac an Ghaill 1996a: 1). Current academic conceptualizations of masculinity should provide more complex understandings of the term through concepts such as 'hegemonic' and 'multiple' masculinities (Mac an Ghaill 1996a: 2). Indeed, 'masculinities are problematic, negotiated and contested within frameworks at the individual, organizational, cultural and societal levels' (Mac an Ghaill 1996a: 2). In this chapter we present a specific focus on one set of familial relations, acting at individual, social and cultural levels, in which masculinities are at play and potentially in great tension with each other, those between fathers and sons.

Connell emphasizes the heterogeneity of masculinity through a recognition of the relations between different kinds of masculinity, which he names 'alliance, dominance and subordination'. Practices of exclusion, inclusion, intimidation and exploitation are part of the gender politics of masculinity (1995: 37). An example of exclusionary, intimidating and exploitative masculinity is evident in homophobia, which Connell defines as a real social practice that creates oppression (1995: 40). While masculinity is a place in male and female gender relations there are gender relations of subordination between groups of men. Connell argues that 'gay men are subordinated to straight men by an array of quite material practices' (1995: 78). We argue that one of these is the ways in which fathers react to the news that their son is gay. Connell explores a set of case studies of gay men within a chapter titled 'A very straight gay' which he opens with a powerful statement: 'No relationship among men in the contemporary Western world carries more symbolic freight than the one between straight and gay' (1995: 143). In all of his case studies (as in our own cases) the men have been, and continue to be, exposed to, and interact with, hegemonic masculinity but through a discovery of their homosexual sexual identities they have transformed the way in which they engage with such a hegemony. The men are both marginalized from, and subordinated by, the dominant masculinity. Nevertheless some of them resist the negative practices and reject the 'patriarchal dividend' because they constitute a masculinity which contradicts the heterosexual masculine hegemony.

Lynne Segal (1997: xxxi) claims that gay men, when collectively organized, have done a great deal to connect theoretical debates on the insecurity of masculinity with a practice of sexual politics which aims to undermine and challenge it. She argues that it is not so much women who make men insecure and fearful but other men, in particular gay men who constitute 'traitors to the cause' (1997: 135). Segal explains why homophobia is a significant part of powerful masculinity. Not wanting to be the object of other men's desires is 'perhaps the most immediate, concrete and consistent proof many men feel they have of their own masculinity' (1997: 134). Contemporary heterosexual masculinity is very dependent on keeping a distance from the opposing category, the homosexual man. Keeping a distance is often translated into an 'obsessive denunciation' of gay male identities which can be as extreme as physical violence.

Consequently, masculinity is a gender project which is part of gender practices that take place through an individual life course, through symbolic practices and in sites of gender configuration, one of which we name as the 'family'.[2] It is also about homophobia and rejection of the 'other', the gay man. For the young gay men in our case studies, a significant moment in their individual life course to date has been the process of 'coming out' to their parents. This process has been part of a symbolic practice in the construction of their masculine identity and has placed them in a potentially conflictual relationship with their heterosexual fathers who are engaged in a different gender project. The site where this all takes place is in the family, so let us turn to notions of fatherhood (see Marsiglio 1995b).

Fatherhood can be described as one of the key signifiers of masculinity. For many men becoming a father is an important part of their masculine identity, part of their 'gender project' (Connell 1995: 72). Fatherhood is usually taken as proof of heterosexuality (although of course significant numbers of gay men father children). It is a powerful symbol, reinforced through popular culture in diverse ways, and contributes to 'hegemonic masculinity', which is bound up with homophobia. The power of the symbol is evident through the current academic, cultural and institutional emphasis on fathering as 'another way of asserting the importance of the traditional *heterosexual* nuclear family' (Segal 1997: 53, our emphasis).[3]

Since the late 1980s, and beyond, there has been an increasing interest in the concept of fatherhood, the practice of fathering as a social rather than a purely biological activity, and a growing discourse around the 'new father' (Lewis and O'Brien 1987; Marsiglio 1995a; Segal 1997). Within geography Stuart Aitken is one of the few to have explicitly focused on 'fatherhood' (1998, 2000, although see Valentine 2004). Fatherhood and the social practice of fathering are usually linked with the socio-spatial institution of the family and in many cases this takes place within the specific place of the household.[4] Some argue that the family is the 'first site in which masculinities are constructed. While fathers are the first role models for their sons' masculinities and take an active part in shaping their sons' construction of masculinity, outcomes are problematic, negotiated and contested' (Heward 1996: 37). Hence, in many, but not all cases, fatherhood, masculinity and family are linked. This interlinkage is important in the relationship between gay sons and heterosexual fathers because it can be the space in which fatherhood and masculinity coincide to perpetuate forms of homophobia.

The concept of the 'new father' has opened up the space in which diverse father-hoods can be discussed. 'Over the last two hundred years the "dominant motif" of fatherhood has shifted from "moral teacher" to "breadwinner" to "sex role model" to "new nurturant father"' (Daly 1995: 22). The 'new father' is defined as a man who is 'highly nurturant towards his children and increasingly involved in their care and the housework' (Lewis and O'Brien 1987: 1). Hawkins *et al.* (1995), discuss the role caring fatherhood can play in men's 'generativity' (care for the next genera-tion), which they argue is 'essential to adults' psychological development' (p. 45). The 'new man/new father' is no longer the distant income earner but a man who co-parents his children and is now much more emotionally involved with them (Aitken 2000; Daly 1995; Westwood 1996). There are tensions within the liter-ature on fatherhood between those that refute or are sceptical about the idea of the 'new father' (Lewis and O'Brien 1987; Segal 1997; Westwood 1996) and those that suggest there are significant changes in the practice of fathering and that alternative fatherhoods need to be better recognized and conceptualized (Aitken 2000). Never-theless, while the 'culture' of fatherhood, the social construction and shared beliefs about men's parenting have changed quite dramatically, the 'conduct' of fatherhood, what men actually do as fathers, has been much more slow to change (Daly 1995; La Rossa 1988; Segal 1997; Valentine 2004; Westwood 1996). In reality children's principal carer and the person who invests most in their children's emotional upbringing is usually their mother (Hawkins *et al.* 1995; Heward 1996; Lewis and O'Brien 1987; Valentine 2004). The lack of change in the 'conduct' of fatherhood is very evident in father's reactions to their sons coming out as gay. The culture of the new nurturant, caring father appears to be absent when hegemonic, heterosexual masculinity is challenged.

Masculinity and fatherhood are highly contested and heterogeneous identities. We cannot talk about *one* masculinity nor can we speak of *a* fatherhood. Power structures and differentials exist between groups of men, and this includes between fathers and sons. A significant feature of masculine power is homophobia. For many men this will be translated into fatherhood through fathering practices, expectations of son's achievements and behaviours and *vis-à-vis* relations with other fathers.

Although there are hegemonic constructions of both masculinities and father-hoods, they are not immune from resistance, challenge and contradiction. Individual experiences of young gay men coming out to their fathers may play a role in the dis-ruption of the notion that masculinity is heterosexual. Hence for a father with a gay son the 'other' masculinity which the father has been encouraged to consistently denounce through hegemonic, homophobic masculinist discourses is standing right before him in the body of his own son. How the father reacts is highly dependent on the ways in which he defines his own masculinity and the depth of his emotional attachment to his son. As we shall see below the clash between the two can be explosive, in other cases supportive. What we present in the rest of this chapter are selected case studies that illustrate the diversity of fatherhood practices and tensions between different forms of masculinity as played out through sexual identities.

The research project

The following case studies are taken from a research project (funded by the Economic and Social Research Council) that focused on the marginalization of young lesbians, gay men and D/deaf people.[5] We worked with 11 young lesbians and nine gay men aged 16–25, five of whom were D/deaf. We also conducted retrospective interviews with 11 lesbians and 13 gay men over the age of 25, six of whom were D/deaf. Fifteen young D/deaf, straight people were interviewed and 15 D/deaf people over the age of 25. Twenty-nine service providers were interviewed and 1,177 school and college students aged 16–18 completed questionnaires. A range of articles and chapters has been published from the research, some of which focus explicitly on the lesbian and gay men related research (Valentine *et al.* 2002, 2003).

For the purposes of this chapter we have selected three of the young, hearing, gay men's narratives to focus upon: Robbie's, Mark's and Noel's.[6] These young men articulate their emotions and experiences of coming out and specifically discuss what this meant in terms of their relationships with their fathers.[7] Fatherhood and masculinity were not explicit focuses of the research project but the experience of 'coming out' and other aspects of young gay men's sexual identity was a focus. For these young men a significant feature of coming out as gay in their families was the roles played by their fathers. From these ethnographic, qualitative examples (much like those presented by Connell 1995) we are able to explore some of the contradictions and negotiations of masculinity alongside the heterogeneity of fatherhood practices that take place within particular families.

Case studies of young gay men's coming out to heterosexual fathers

Robbie's story: fatherhood as condemnation and denial

Robbie is a 22-year-old graduate student who is currently employed as a sessional outreach worker for gay and bisexual men's health. He grew up with his mother and father (now both retired) and a brother who is three years older. His sister is 22 years older than him and had left home by the time he was born. Robbie described his parents as conservative, old in their ways and very strict. He and his brother became very involved in Evangelicalism during their teenage years. At University Robbie joined the Christian Union but found that he was simultaneously identifying as gay. When he came out as gay he was forced to give up his involvement in the CU.

Robbie became increasingly unhappy at university and felt he needed to talk things through with someone close to him. He decided to come out to his brother who had already made it clear that he believed being gay was not incompatible with being a Christian. Robbie described his brother as being fine with Robbie's gayness and very supportive. As this had worked out quite well and he really felt the need for support from his parents as he was having such a difficult time, Robbie decided to go home for the weekend and tell his parents.

> It was a complete disaster . . . really horrible. I went home and told them straight away that there was something really important that I wanted to tell them but I didn't want to do it that night, but I had to say that because otherwise I would chicken out over the weekend. I went into town with my Mum and she was really upset and worried and stuff so I told her and she was just . . . really insulting, really upset, went a bit hysterical.

Robbie explained that he knew that his parents might be conservative in their reaction but he trusted in the fact that as they were his parents they would try to understand and listen to him. He admitted that he wildly misjudged his parents' potential reaction.

> We got home and she told Dad and he was like worse because he was like all rational and didn't shout at all about but just came down and said . . . well, he basically said that it was the same thing as being a paedophile and never to tell my sister because if I did she wouldn't let me see my niece anymore and they would support her in that completely cos they'd see me as a really bad influence on her. He said I wouldn't be able to get a job and that my life would be in the gutter but it was my choice.

Robbie stated that although things had gone so badly he felt a sense of relief after telling them because he had always wanted an open relationship with his parents and not one based on deceit and secrets. He had hoped they would be able to discuss the issue more at a later date when they had got used to the idea that he was gay. However,

> They never talked to me about it again apart from one row that we had. I decided to take a year out after my second year and go and live in a monastery because things weren't going that well for me or my degree and I wanted to sort myself out. So I told them and my father went mad about that as well, because I was thinking about getting ordained into the Anglican Church. Dad brought up all the stuff like how can you think of being ordained in the church, you're an abomination!

This extreme condemnation of his son as a 'paedophile' and an 'abomination' is an explicit example of a fatherhood built upon a masculinity which is bound up with homophobia. He does not react to his son's difficulties with the compassion of a father but with the abusive distancing of a homophobic man. Robbie is the very embodiment of the 'other' which the father has clearly rejected in order to construct his own masculine identity. The next episode in the coming out process for

Robbie revolves around the heterosexual institution of marriage and his brother's wedding. Robbie and his father were on formalized speaking-terms as long as sexuality and Robbie's personal life were not the subject of conversation. However, problems arose when Robbie's brother was organizing his wedding. Robbie had been in a relationship with 'Philip' for two years. His brother and fiancé had met Philip several times, enjoyed his company and wanted him to be at the wedding. Robbie's parents knew of Philip's existence but had never mentioned or met him.

> Dad was on the phone to me and he said 'how you getting down for the wedding?' I told him Philip was giving me a lift down. He said, 'but he's not coming to the wedding is he?' and I said yes he was and he said 'but it's a real problem' and I said, well I don't understand, then he wouldn't speak to me and hung up. I found out that the next day my Mum phoned my brother crying and said if Philip went to the wedding Dad wouldn't go and she wouldn't go. My brother tried to explain that I felt the same way about Philip as he did about Jane but she said 'I don't want to hear it because it makes me sick'. So Philip didn't end up coming but they never spoke to me about it, they never even said thank you for not bringing him, we really appreciate it. Now they even pretend that I haven't come out to them, they tell me they knew I was depressed and that they wished I'd talked to them about it! They deny saying any of the things they said to me when I did come out!

Robbie's father has built up a complex web of separation between himself and his son. As a man he constructs a masculinity and fatherhood based on homophobia in relation to his gay son and yet as a father he is totally accepting of his other son. His other son has demonstrated his heterosexuality through marriage and hence a conformity to hegemonic masculinity. Robbie, however, disrupts and challenges this particular construction of masculinity (Segal 1997). There was even a danger that such a challenge would be publicly visible to other men if Robbie and Philip had both attended the wedding. In threatening to disrupt the wedding by not attending, Robbie's father asserted his patriarchal authority and forced his other son and fiancé to make a very difficult emotional decision. Robbie, out of understanding for his brother, agreed not to take Philip. The father forced a homophobic collusion from a brother who in fact is supportive of his gay brother, but loyalty to the father proved stronger.

The denial and silence around Robbie's gay identity by his father may appear to be less confrontational than the fights and arguments Mark and Noel went through but in reality it is extremely controlling. The father effectively re-writes the story and maintains a façade that Robbie is not gay and consequently not a threat or challenge to his own hegemonic masculinity. Robbie feels very hurt and confused about his father's continued denial and non-communication. He sees his father less and less and can foresee a time when they will lose all contact. Although this upsets him greatly he cannot see any way around the situation created by his father.

Mark's story: fatherhood as rejecting and condemning masculinity moving to a degree of fatherly care

Mark is 19 and employed in a private security control room. He grew up in a household with an older sister and his mother and father. He described his relationship with his parents as close. His mother was a cleaner and his father, after a period of redundancy from mining, was a delivery driver. Mark had a strong sense of 'being different' from his peers from the age of 11 and experienced homophobic bullying throughout most of his secondary school career. He had moved out of the family home a few months before the interview.

When he was about 17, Mark was finding it increasingly hard to live a lie with his parents as they had always raised him to be honest. He did some research, gathering literature and details of support groups who helped parents with gay children. His main concern was that in coming out he would hurt and disappoint his parents. Mark had become sexually active and due to extremely poor sexual education at school was concerned that he had not been practising safe sex well enough and that he had contracted HIV. He decided to go for an anonymous test. It was going for this test which proved to be a 'critical moment' (see Giddens 1991; Thomson *et al.* 2002; Valentine *et al.* 2003 for their analysis of this concept) in Mark's relationship with his parents, and particularly his father.

Mark took some time off work to go to the test and so encountered his father at home. His father asked him what he was doing at home and Mark was evasive which aroused suspicion in his father. A few days later Mark's father pushed him further and Mark admitted that he had been to hospital for some tests. He describes his father as getting worried about him and asking more direct questions. Mark admitted that he had been for an HIV test, then:

> his immediate reaction was, 'are you doing drugs?' I was like 'no' . . . 'have you been shagging a bird?' . . . I'm like 'no' . . . then his face turned a very nice shade of white and (he) went all wide eyed and everything, 'have you been with men'? . . . 'Er yes' . . . so he immediately found out I was queer and decided that I was going to die of AIDS, you know, all in the same instant, which was a bit of a shock really.
>
> [. . .]
> 'So what happened?'
> 'He sat there in shock knowing not what to do or what to say, and this is the really hurtful part . . . I was trying to talk to him about it and . . . he called, in the conversation, he managed to call me a pervert, said that queer bashing was normal and that any normal father would have thrown his kid out on the streets for being gay and I just couldn't believe it that my own dad said that, I thought, I knew he might react badly but . . . I was unbelievably hurt by what he'd said . . . I was really upset cos' I didn't want to hurt anyone, I didn't want myself to get hurt.'

Mark left the house and went to his sister's who he had come out to a few weeks before. Mark and his sister then called their paternal grandmother and told her what had happened and she tried to speak with Mark's father. However, her intervention made things worse. They intercepted Mark's mother on her way home from work and Mark told her. She was upset but claimed to have thought he was gay for a while but still was not prepared for the reality. Mark's mother spent some time with his father and later returned to Mark in his sister's home and assured him that she still loved him and that she would be on his side if 'things got nasty'. However, a day or two later 'for the first time ever in our family history . . . we had a family meeting, you know, my dad called this family meeting for me, my mum and him to sit down'.

The meeting was not a chance to talk things through together as Mark had hoped but a space for his father to lay down how Mark was going to live his life in *their* home:

> In the blink of an eye take away all my freedom . . . and liberties . . . they just told me how it was gonna be . . . they laid down these five golden rules . . . I wasn't to flaunt my homosexuality in the house . . . I wasn't to talk about that side of my life . . . I wasn't to bring any boyfriends home . . . I wasn't to bring any of my friends who were 'that way' home . . . they decided I couldn't tell anybody else.

Mark tried to reason with his mother:

> but my dad just didn't listen, the more I pressed the subject the more angry he got and I knew that if I pushed the matter too far that they would throw me out of the house and you know they were saying to me that it seemed as though I wasn't their son anymore, that I'd been replaced by someone else, that I was a stranger to them, and the weird part was, I felt the same as well, I mean I sat there looking at and talking to these people and I didn't see my mum and dad anymore, I didn't see these two people that loved and cared for me, regardless, and who brought me up and everything [I just saw] these two strangers, these authoritarian people . . . taking away all my freedom and liberty and human rights and everything and I was flabbergasted . . . [Then] my dad just tried to make everything go along as normal as though nothing had happened.

For some time things were extremely strained. Mark's mum began to soften in her attitude and would sometimes mention his boyfriend and stress that she wanted him to be happy. His current partner, who Mark was in process of setting up home with at the time of the interview, is known about by Mark's parents but his father initially made it explicit that Mark's partner was not allowed in the family home. At the time of the interview, however, Mark's father had met his boyfriend a couple of times, had invited him to the New Year's Eve party and was helping them move into their own place. It had taken Mark's father two years to reach this level of 're-acceptance' of his son.

From Mark's story we gain an insight into a fatherhood which is closely tied to a heterosexual masculinity based upon a total rejection of a homosexual masculinity (Connell 1995). Mark's father rejected Mark as the wrong kind of man and his homophobia overrode his fatherhood of, and emotional attachment to, Mark. His father re-established his 'rightful' place as the family patriarch (over-riding Mark's mother's wishes) and laid down family laws based upon a singular type of masculinity. There was no room in the house for any evidence of an alternative masculinity (Segal 1997). Mark's father argued effectively that he could not see Mark as his son anymore just as Mark could not believe his own father would treat him that way. Time and interventions by Mark's mother and his sister have provided the space for some degree of shift in the father's hegemonic heterosexual masculinity. Mark's current boyfriend has been allowed into the patriarchal home for a visit. Mark's father was slowly relaxing his exertion of a masculine privilege and allowing his caring fatherhood of Mark to re-emerge albeit slowly. Mark remained hopeful that he can continue to rebuild a positive relationship with his father but knew his father would never fully come to terms with his son's gayness.

Noel's story: fatherhood as rejection, acceptance and formation of a new father–son relationship

Noel had just turned 21 at the time of the interview and had very recently lost his job. He described his teenage years as reclusive and his behaviour worried his mother so much that she had taken him to a child psychologist. However, attending college 'prized me out of my shell' and he got into drugs, partying and experimenting. His father and mother separated when he was young. His mother remarried and so Noel lived with his mother and step (social) father and one sister throughout his childhood (his other two sisters lived with their biological father). His mother was a cleaner and his stepfather (as Noel refers to him) a window cleaner. He said that he was not in touch with his father during his childhood. At the time of the interview he was living with his mother and stepfather but was planning to move in with his boyfriend at the end of the month.

Noel came out to friends at college about ten months before telling his parents. Noel said that he didn't plan to come out to his parents as such but 'blurted it out during an argument'. He said that they did not believe him and assumed that because he had had some medical problems with his penis that it was anxiety around that which had made him think he was gay. Later they asked him directly if he had been with another man, and he had lied and said 'no' but a while later he told them the truth that he had been sexually active with a man. He said his parents then presumed that he had been pressured into having sex and that it was really a phase he was going through. However, when Noel got a steady boyfriend in his life it made his gayness a reality.

Once it became a reality he said it was all quite confusing in terms of his parents' reaction:

> some days it'd be they'd talk about it for ages and ages and other days it'd be 'don't bring shame on our house' . . . I just didn't know. At times my mum would seem very understanding about it and then other days she would make some silly glib comments about it . . . then my step-dad he would, I'd talk to him about it and think 'oh I've won him over' and then he'd just say something absolutely ridiculous so I just never knew where I stood with them.

Sometimes he described it as quite funny as his stepfather would use the term 'queer' as a figure of speech for something that wasn't quite right and his mother would tell him not to say it, 'then of course, my dad, of all people, would go on about gay rock stars a lot!' Noel also recalls how they used to tell him how life would be hard for him and how lucky he was that they had not thrown him onto the street and had accepted it. Mark's sister though was extremely hostile about Noel's boyfriend and his parents blamed Noel for being 'indiscreet', accusing him of abusing their trust as they had allowed his boyfriend to visit and stay the night in Mark's room. For a while, because of his sister's hostility, Noel was not allowed to have his boyfriend to stay and so they resorted to driving in his car and 'getting into it in bloody lay-bys!'.

What is clear from this part of Noel's coming out story is that his stepfather reacted in a rather bemused and confused way. Initially neither he nor his wife believed what Noel told them, they assumed it was a phase, but his stepfather did not have an immediate negative reaction. The person who reacted most badly was Noel's sister but his parents chose to take her side and changed the 'house rules' which had initially been open and tolerant. However, a subsequent stage in family relations took place and Noel's biological father became involved.

Noel came home drunk one night, had a row with his mother who was also drunk and she slapped him across the face. He tried to retaliate and his stepfather intervened and 'basically we started beating each other up and trashing the house'. Noel's mother threatened to call Noel's 'real dad' who is a 'big, great big man'. Noel carried on fighting and arguing and so his 'real dad' was called. His biological father (hereafter called 'father') came round and pinned Noel down and found out about Noel needing to have an operation and also about him having come out as gay.

> He was very, very good about it, said it didn't matter to him . . . he come round the day after and we had conversations about my operation, we talked, we even talked about sex, about safe sex which was something my parents, my Mum and my step-dad never would have dreamed of talking

about, as far as they were concerned they didn't wanna know . . . they can handle a few things but they said they don't wanna know about that side of my life . . . my dad (father) . . . I was a bit surprised . . . he did have a discussion with me on safer sex . . . and says 'Oh, if I ever meet [name of boyfriend] what can I call him? Your special friend?' . . . it meant a lot to me that my dad who I barely knew was so supportive, he was a lot more supportive to my mum as well so it was a blessing in disguise that fight really cos' my dad has been and still, has been quite supportive since, I've started to get to know him better now . . . since he found out.

A little later in the interview Noel said that he had found that his mother seemed to have more problems with his sexuality than his stepfather did. He described his stepfather as being 'nonplussed' and that he got on very well with Noel's boyfriend, 'he doesn't judge people'. Noel comments on an irony that in his boyfriend's family the father had taken it much worse than the mother but that:

it was the women in my family that took it worst, yet all the men, all seem, were all very good about it you know, now I can have discussions about my sexuality and my gay friends and everything with my step-dad as if I was just talking to one of my mates . . . its definitely improved my relationship with my step-dad anyway.

In many ways Noel's experience with both of his fathers is the antithesis to Robbie's and Mark's fathers' reactions. Noel's stepfather has made a transition from confusion and bemusement to one of open tolerance and a genuine interest about Noel's gay life. There might be detailed information that he does not want to hear, but that is probably true of most fathers with their children. A critical moment of violence and genuine anger and frustration in Noel's life resulted in a space in which to build a new relationship with a father who had been largely absent from his life. Noel's father's acceptance of his sexuality and his open support has been extremely important for Noel. Both fathers in Noel's life have resisted a retreat into hegemonic heterosexual masculinity and have worked through a way to be able to engage with Noel's different/other masculine identity.

Conclusion

From these case studies it is clear that there are various ways in which fatherhoods and masculinities are mapped out in the socio-spatial relations of the family. As gay sons came out to their parents (biological or social) choices were made by their fathers about how to react and what fatherhood/masculinity position to adopt. In Robbie's and Mark's cases hegemonic masculinities and patriarchal fatherhoods were constructed; in Noel's varying examples of gentler and more tolerant fatherly and masculine reactions were established. Robbie feels that there has been an irreparable separation between him and his father; his father chooses distances from the 'homo-sexual other' (Segal 1997). Mark remains hopeful but sceptical about his relationship with his own father, but at least there is the space for some shift in the father's mas-culine identity (Connell 1995; Mac an Ghaill 1996a). Noel feels much more com-fortable with both his fathers and their improved relationship (Daly 1995). We would argue that Robbie's and Mark's fathers would appear, in the masculinist con-struction of fatherhood, to have maintained the powerful hegemonic position, but the cost to both father and son has been painful and is perhaps irreparable. It is clear that masculine homophobia damages important familial relationships, whereas fatherhoods built upon caring and nurturing masculinities create the space for nego-tiation, acceptance and rewarding relationships.

Notes

1 We problematize the notion of 'family' because we recognize that families do not always constitute the 'hegemonic structure' of mother, father and their children, the nuclear family. Household structures are far more complicated than this and we have to be wary of the term 'family' because it often hides the complexity and diversity of this particular social institution.

2 Interestingly in his discussion of gender projects and practices Connell names just three sites of gender configuration: the state, the workplace and the school (1995: 72–3). We argue that the family is a significant omission from this list but that this omission parallels what Aitken has termed Connell's muteness in relation to fatherhood (Aitken 2000: 583).

3 We would insert here 'so-called traditional heterosexual nuclear family' because there is considerable evidence within family studies to show that the nuclear family has never really been predominant, its 'traditional' status is largely ideological.

4 We recognize that fathering can take place outside of the family home/household, that offspring may not reside with their biological fathers and that step or social fathers often play an important role in the lives of children who are not biologically their own. We also recognize that fathering is something that takes place over a long period of time, on into the adult lives of sons and daughters. However, there appears to be minimal discussion of the role of adult fathering of adult children, with the exception of feminist explorations of father–daughter relationships.

5 D/deaf is written in this way to indicate the political and cultural differences within D/deaf communities between those who consider themselves to be a cultural and linguis-tic minority (Deaf people) and those who identify as having a disability (deaf people). For a more complex discussion of these identities see Skelton and Valentine (2003a, 2003b).

6 All of these names are pseudonyms to protect the young men's anonymity. Where we use direct quotations these are as the young men spoke them, hence grammatical errors are part and parcel of their Midlands expressions and dialects.

7 Other young gay men talked about their coming out experiences and named their fathers in specific ways but these three provide an insight into different fatherhood reactions to having a gay son.

References

Aitken, S. (1998) *Family Fantasies and Community Space*, New Brunswick, NJ: Rutgers University Press.

Aitken, S. (2000) 'Fathering and faltering: "sorry, but you don't have the necessary accoutrements"', *Environment and Planning A* 32: 581–98.

Connell, R.W. (1995) *Masculinities*, Cambridge: Polity Press.

Daly, K.J. (1995) 'Reshaping fatherhood: finding the models', in Marsiglio, W. (ed.) *Fatherhood: Contemporary Theory, Research, and Social Policy*, Thousand Oaks, Ca: Sage Publications.

Giddens, A. (1991) *Modernity and Self-identity: Self and Society in the Late Modern Age*, Cambridge: Polity Press.

Hawkins, A.J., Christiansen, S.L., Sargent, K.P. and Hill, E.J. (1995) 'Rethinking fathers' involvement in child care: a developmental perspective', in Marsiglio, W. (ed.) *Fatherhood: Contemporary Theory, Research, and Social Policy*, Thousand Oaks, Ca: Sage Publications.

Heward, C. (1996) 'Masculinities and families', in Mac an Ghaill, M. (ed.) *Understanding Masculinities: Social Relations and Cultural Arenas*, Buckingham, Open University Press.

La Rossa, R. (1988) Fatherhood and social change, *Family Relations* 36: 451–8.

Lewis, C. and O'Brien, M. (1987) 'Constraints on fathers: research, theory and clinical practice', in Lewis, C. and O'Brien, M. (eds) *Reassessing Fatherhood: New Observations on Fathers and the Modern Family*, London: Sage Publications.

Longhurst, R. (2000) 'Geography and gender: masculinities, male identity and men', *Progress in Human Geography* 24(3): 439–44.

Mac an Ghaill, M. (1996a) 'Introduction', in Mac an Ghaill, M. (ed.) *Understanding Masculinities: Social Relations and Cultural Arenas*, Buckingham: Open University Press.

Mac an Ghaill, M. (ed.) (1996b) *Understanding Masculinities: Social Relations and Cultural Arenas*, Buckingham: Open University Press.

Marsiglio, W. (1995a) 'Fatherhood scholarship: an overview and agenda for the future', in Marsiglio, W. (ed.) *Fatherhood: contemporary theory, research, and social policy*, Thousand Oaks, Ca: Sage Publications.

Marsiglio, W. (ed.) (1995b) *Fatherhood: Contemporary Theory, Research and Social Policy*, Thousand Oaks, Ca: Sage Publications.

Segal, L. (1997) *Slow Motion: Updated and Revised* (2nd edn), London: Virago.

Skelton, T. and Valentine, G. (2003a) 'Political participation, political action and political identities', *Space and Polity* 7(2): 117–34.

Skelton, T. and Valentine, G. (2003b) 'It feels like being Deaf is normal: an exploration into the complexities of defining D/deafness and young people's D/deaf identities', *The Canadian Geographer*, (forthcoming): n.p.

Thomson, R., Bell, R., Holland, J., Henderson, S., McGrellis, S. and Sharpe, S. (2002) 'Critical moments: choices, chance and opportunity in young people's narratives of transition', *Sociology* 36: 335–54.

Valentine, G. (2004) *The Crisis of Childhood: Children and Production of Public Space*, Aldershot: Ashgate.

Valentine, G., Skelton, T. and Butler, R. (2002) 'The vulnerability and marginalisation of lesbian and gay youth', *Youth and Policy: The Journal of Critical Analysis* 75: 4–29.

Valentine, G., Skelton, T. and Butler, R. (2003) 'Coming out and out-comes: negotiating lesbian and gay identities with, and in, the family', *Environment & Planning D: Society and Space* 21(4): 479–99.

Westwood, S. (1996) 'Feckless fathers: masculinities and the British state', in Mac an Ghaill, M. (ed.) *Understanding Masculinities: Social Relations and Cultural Arenas*, Buckingham: Open University Press.

Further reading

Connell, R.W. (1995) *Masculinities*, Cambridge: Polity Press. This is one of the definitive texts on the subject of masculinity and it is written in an engaging and readable style. The case study material is extremely interesting. The text is equally valuable for students and researchers.

Mac an Ghaill, M. (ed.) (1996) *Understanding Masculinities: Social Relations and Cultural Arenas*, Buckingham: Open University Press. This rich edited collection brings together diverse analyses of masculinity written by some of the leading writers in the field. The essays are thought provoking and based on a range of research which enables the reader to engage with a diversity of masculinities.

Marsiglio, W. (ed.) (1995) *Fatherhood: Contemporary Theory, Research, and Social Policy*, Thousand Oaks, Ca: Sage Publications. This book provides a useful collection of essays on a wide range of debates relating to fatherhood through time and space. It is rather US-centric but provides a useful context for comparison with discussions about fatherhood in different geographical spaces.

Segal, L. (1997) *Slow Motion: Updated and Revised* (2nd edn), London: Virago. This is an extremely readable and provocative book. It provides an insight into a wide range of masculinities from a feminist position. Although there have been some political and legal changes since the publication of this book, it nevertheless provides an important insight into why social constructions of masculinity still have to change considerably for a more equal society.

17

THE AWKWARD SPACES OF FATHERING

Stuart C. Aitken

Summary

This chapter takes issue with the notion of fatherhood evolving from the distant breadwinner of the nineteenth century, through the genial dad and sex role model of most of the twentieth century to today's father as equal co-parent. I argue that the work of fathering is different from that of mothering, and that there is little theory or empirical data as yet to help us understand how space is produced and time contextualized and gendered by this kind of work. A central problem of the chapter is to highlight how much of the institution of fatherhood does not embrace fathering as a daily emotional practice that is negotiated, contested and resisted differently in different spaces.

- A labour of love
- Excluded geographies of fathering
- The culture and conduct of fathering: power shifts and awkward spaces
- Fathers and family space: a concluding comment.

Introduction

During the nineteenth century, William Ewart Gladstone famously served as prime minister for Great Britain on three occasions and is known for his spirited rivalry with Disraeli, his attempts to rehabilitate prostitutes, his public educational reforms and his championing of home rule for Ireland. He entered public office in 1832 and served there until 1895, except for a one-year absence. What is almost never included in his biographies, or is included only as a short footnote, is that Gladstone removed himself from public office to spend a protracted period of time sitting at the bed-side of his terminally ill infant daughter Catherine, who died in 1850. Gladstone the father is hidden from his lauded public history because masculinity, as a symbolic and mythic form of identity, is overtly determined for its coherence upon external public discourse (Laqueur 1992). Fathering histories remain hidden because they are awkward, rarely fitting the geography of public ventures and power. The argument of this chapter follows, then, that the geographies of man-as-father are almost exclusively subsumed under the monolithic geography of a pervasive patriarchy that includes the space of public authority and its transmission over generations. What is missing is the somewhat problematic notion of the public-man-in-private, which nonetheless suggests the elaboration of both geographies and histories of fathering.

As a consequence, and as part of a seeming private sphere, fathers confront an identity predicament that is often hidden, and is always awkward. The awkwardness stems, in part, from unclear models on how men should be fathers as well as a lack of recognition of what constitutes the work of fathering. That awkwardness is highlighted with increased public employment for women over the last few decades that suggests a shift in familial power relations. Whether the patriarchal nuclear family is waning or simply shifting face, it is nonetheless important to study the consequences of these kinds of cultural transformations for fathers. For, it is entirely likely that without discussion of the consequences of change, without discussion of appropriate and inappropriate signposts and models, the awkward spaces of fathering will grow.

I seek to do three things with this chapter. First, I want to highlight the emotional work of fathering that takes us beyond mythic ideas of fatherhood. Second, I want to move the body of knowledge on masculinities forward with a focus on fathers that specifically addresses how masculine identities are learnt in relation to women and children, and shaped by produced and occupied family and community spaces. Third, I want to move our knowledge of fathering beyond its definition in opposition to mothering. These issues are complex and contradictory. My overall concern here is to focus on connectedness and relationships rather than parallelisms and discontinuities. My enduring presumption is that many academic studies and policy proclamations slip into a false parallelism between mothers and fathers. This position upholds fatherhood ideology and research that defines a father's relationship and involvement with his children primarily as a form of co-parenting that is interdependent with, in opposition to, and at times less than mothering. It is a position that often soft-pedals emotional labour and the connectedness between fathers and their children. I propose that a set of mythic ideals help structure the gender and spatial relations of fathering, and these often get in the way of day-to-day work of fathering. A central problem of contemporary work, I argue, should be to highlight how much of the institution of fatherhood hinges on an 'idea' that does not embrace the 'fact' of fathering as a daily emotional practice that is negotiated, contested and resisted differently in different spaces.

I begin with a labour theory of value that moves beyond essentialized notions of mothers as more natural parents and focuses instead on the work of parenting. I then elaborate spatial power structures by tracing transformations of the idea of fatherhood, and ruminate on the ways that it is situated and embodied politically. I end the chapter with a discussion of the spaces of fathering and elaborate a need to incorporate the emotional work of fathering into discourses on masculinity.

A labour of love

To begin, then, I highlight the emotional work of fathering by focusing upon the value of parenting as labour that is emotionally charged as opposed to biologically determined. In an impressive and not-well-known essay, historian Thomas Laqueur (1992) forcefully argues that the 'biological facts' of motherhood and fatherhood are not 'given' but come into being as science progresses and identities assume cultural significance through political struggle. As an extension of his earlier opus on the social and biological construction of sex, Laqueur (1990, 1992: 155) puts forward a 'labour theory of parenthood in which emotional work counts'. The notion of emotional bonds determining rights of parenthood flies in the face of many legislative motions, including a 1999 US Supreme Court decision favoring mothers in child custody and citizenship disputes (*Nguyen v. INS*, 99–2071). The decision to accord mothers more rights than fathers when a child's citizenship is disputed raises larger issues of who is a more 'natural' parent: 'The difference between men and women in relation to the birth process is a real one', said Justice Kennedy concerning this legislative move, 'and it provides a reasonable basis for the law' (*Los Angeles Times*, 12 June 2001, A9). It seems to me that the position of this Supreme Court judge flies in the face of contemporary academic discomfort that immutable qualities exist on the grounds of sexual difference, and that these qualities are natural rather than socially constructed.

If we embrace feminism as a powerful denaturalizing force then we must conclude that gender is a socially imposed division of the sexes, and this includes women's relations to the birth process. As Audrey Kobayashi (1994: 77) points out, we need to fundamentally challenge anything that is meant to characterize 'people according to criteria which may seem to present themselves as natural (because they have been naturalized) but are nonetheless based upon social choices'. But the position of the US Supreme Court that the natural right of motherhood is a reasonable basis for the law, presupposes that motherhood is ontologically different from fatherhood. Proponents of greater rights for fathers, such as Laqueur, argue alternatively that no fact – such as the ability to bear and nurse children – entails or excludes a moral right or commitment. His point is that if this ontological difference between mothers and fathers is a reasonable basis for law, then the *idea* of fatherhood can never triumph over the *fact* of motherhood.

In actuality, Laqueur (1992: 158) goes on to argue, the facts of motherhood – and of fatherhood – are not given only through corporeal bonds but also come into being through political struggles and as scientific and legal ideologies progress. The 'idea' that a child is of a parent's flesh and blood is only a partial embodiment because 'biological correlatives and their cultural importance depend on the available supplies of fact and their interpretation'. Nothing of the US Supreme Court decision speaks to a moral right or commitment, and this is where Laqueur's labour theory of parenthood finds its power. Moral rights and commitments are born from connectedness and are emotive. Laqueur borrows David Hume's argument that such things as wild animals and spectacular landscapes may evoke little sense of obligation unless they belong to, or are connected to, us (Hume 1955). Hume points out that most people fear a minor injury to themselves while caring almost nothing about the death of a distant stranger on the news. The issue for Hume is not the objects, but our relationship to them. Detached and distant objects do not arouse emotions and, further, moral concern and action are engendered not by the logic of the relationship between human beings (e.g. giving birth), 'but by the degree to which emotional and imaginative connections that entail love or obligation have been forged' (Laqueur 1992: 164). Arguments of this kind beg for a fuller understanding of emotions in daily life.

Kay Anderson and Sue Smith (2001: 7) claim that the neglect of emotions in social science research suppresses 'a key set of relations through which lives are lived and societies made'. Emotions are highly political and sexualized but until recently they have rarely been enframed and emblazoned by academics as an important component of public action and responsibility. To paraphrase Anderson and Smith, there are times and places where lives are explicitly lived through pain, love, shame, passion, and anger to the extent that it is hard to discount the ways emotional relations dictate social practices and moral commitments. All relationships are emotionally charged, and the emotional spaces between parents and children are no exception. That said, familial spaces are forged over extended periods of time and they are gelled with commitment and obligation. The biological connections between parents and their children are not just about sperm, gestation, labour, birth, transferred DNA and nursing. Each of these facts is loaded with centuries of cultural evolution co-mingled with biological evolution. As Laqueur (1992: 158–9) points out forcefully, 'laws, customs, and precepts, sentiments, emotions, and the power of the imagination make biological facts assume cultural significance'. The fact of motherhood is not just about the reality of giving birth, it is also about the labour that goes into making connections and appropriating the fetus and baby into the mother's moral and emotional space. I do not want to diminish the importance of the embodied connection between mother and child, but there is clearly an emotional geography at work here for fathers and mothers that transcends, transforms and reifies physical bonds. And so, from the perspective of Laqueur's labour theory of parenting, the fact of fatherhood is also based on the work that goes into emotional and moral connections. This kind of labour theory of value gives mothers and fathers rights to a child.

But to what extent do fathers embrace the emotions of parenting without committing to the responsibility of domestic labour – labour of the hand rather than labour of the heart – that such emotions entail? I raise this question to highlight another side to the story surrounding the US Supreme Court action in 1999. It resonates in problematic ways with current critiques of fatherhood ideology because this kind of decision establishes a loophole for men to disassociate themselves from parenting responsibilities. Frank Furstenberg (1988) provides insight into this conundrum with his celebrated 'good dad – bad dad complex' (see also Biller 1993).

Through the externalization of women's roles and their increased employment, he argues, men can escape from 'good provider' responsibilities. In his book, *Fatherless America*, Blakenhorn (1995) takes this point further, somewhat problematically, to suggest that returning men to responsibility should focus primarily on their financial commitments. Furstenberg's (1988) enduring point is that fatherhood today assumes a voluntary dimension and fathers can either retreat from responsibility or participate more fully in the family, but contemporary 'ideas' of fatherhood and many legal structures enable retreat. In a related critique of the social science literature on parenting, Louise Silverstein (1996: 11) uncovers an ideology that reflects 'the belief that active participation by mothers in the daily care of children is obligatory, whereas nurturing and caretaking by fathers is discretionary'.

Excluded geographies of fathering

This brings me to consider some of the ways fathering is hidden in geography as a discipline and, perhaps more importantly, the ways that the spaces of fathering remain outside the purview of most social science research. Consideration of this lack, I think, suggests an important move forward in the literature on masculinities as a whole. It is important to reflect on how fathering identities are learnt in relation to women and children, and also how they are shaped by the produced and occupied spaces of families and communities.

Although other disciplines across the social sciences have begun to explore the contested terrain of fatherhood, geographers seem reluctant to include issues of fathering in their analysis of space. This is despite frequent calls to analyze (gender) identities relationally and differentially (Massey 1995; Duncan 1996). A focus on the relational formation of fathering identities and masculine spaces seems long overdue in both feminist and gender-oriented geographical work. The work undertaken by Bell and Valentine (1995) on geographies of sexuality says nothing about fathering. Longhurst's (2000) review of geographic research on masculinity highlights the need to move away from essentialists' conception of identity, with an overwhelming focus on challenging heteronormativity and exorcising homophobia, but there is no mention of new ways of representing fathering.

In what follows, I want to paint the geography and history of fathering in broad swathes. What I have to say is not unproblematic, but it enables a quick move to the heart of my concerns with the contemporary uneasy spaces of fathering. A more detailed account of the quirky geographic processes surrounding family formation may be found in my *Family Fantasies* book (Aitken 1998).

By the time Western society achieved industrialization there was a strong spatial component to the evolving political identities of men, women, and children, and they all became encompassed in a transforming social imaginary of the family. Fathers were bequeathed control of resources and productive activities in the public sphere and mothers were seen as having total control and unlimited power with reproductive activities (Hayford 1974; Gottlieb 1993). Being a 'father' became quite different from being a 'mother', and the roles of sons and daughters came to depend upon age, sex, and consanguinity (Bernardes 1985: 281). With the spatial evolution of Western cities through the last century and a half, feminist geographers suggested that the spatial entrapment of mothers and children in suburbia is contrasted with an ever more distant father who maintained control of productive, public space (Fagnani 1993; Hanson and Pratt 1995). This separation of father from the private sphere became important for capital productivity, so too did the isolation of mothers and children from that same productivity. It established a mythic image of the ideal mother who would guarantee both morally perfect children and a morally desirable society (Bloch 1978; Marsh 1990) from within the private sphere, and it enabled fathers to disengage emotionally from that sphere. This is not to suggest that fathers loved their families less, it simply provided a loophole for moral disengagement. This inconsistency needs further exploration as it bears heavily on my argument for the need to refocus on the work of fathering as an emotional act rather than on fatherhood as some mythic institution.

And so, importantly, a series of paradoxes arise because this emotional separation and individuation through the late nineteenth and into the twentieth century does not reflect day-to-day life in the same way that it buttresses society's mythic ideals of social reproduction. There are some interesting consequences to this inconsistency. For example, after some struggle through various factory and labour laws in Europe and the US, the place of children and women was squarely set in the private sphere. Of course, in reality, women were not always and in all places at home with children. The factory and labour laws of this time removed some women from workplaces or reduced the number of hours they could work to the extent that it may be argued they were not pushed into the private sphere with unlimited power over reproductive activities but they were, rather, infantilized and made powerless with their children. Paradoxically, by the second half of the twentieth century, when women again entered the paid labour force in significant numbers and some chose not to become mothers, they were blamed not only for maternal deprivation but also for overprotection of children. A dominant ideology surfaced and joined with Freudian psychological theory to blame mothers for any failings in children, even if they had no children.

At the same time, Janice Drakish (1989: 73) suggests 'social science research is constructing fatherhood as a panacea for the problems mothers created and cannot control'. There are a number of examples in child development research that assume that a father's presence is almost always a good thing (e.g. Lewis *et al.* 1981; Palkovitz 1997; Lamb 1997). For the most part, this literature points to the father as the co-parent who is responsible for gender, cognitive, intellectual and academic evolvement. Hawkins and Dollahite (1997) go further, introducing the term 'generative fathering' to argue that previous research on fathers is flawed because it fails to accommodate men's desire and ability to care for the next generation. Generative fathering, in their view, can and will emerge as a dominant image of fatherhood if researchers stop focusing on men's sociohistorical move away from the responsibilities of the family (see also Hawkins and Palkovitz 1999). Although there is some merit to this perspective, there is danger in suggesting that men's roles as fathers revolve around the transmission of culture and over-seeing rites of passage. In an essay on parenting as a right of passage that focuses on the way that young men and women learn their roles, I suggest that patriarchal bargains are also transferred as a covert cultural norm (Aitken 1999). Gender role distinctions of this kind are clearly problematic for their eliding of power relations and their disregard of how political identities are created and transformed in and through space. In actuality, the day-to-day work of fathering is almost completely hidden from history and its geography is highlighted only in as much as a spatial separation of the public and private sphere is deemed important.

That many social science studies make no attempt to problematize the work of diverse types of fathering suggests that this form of male power is still regarded as a backwater to the dominant histories and geographies of male public power. Indeed, it may be argued that notions such as 'generative fathering' are less about the care of children and more about men's public authority and its transmission over generations. What is clear from many of these contemporary academic discussions about fathering is that their roots are changes in the ideology and structure of family life and that these changes precipitate men into a state of uncertainty in their relationship to the private sphere.

The culture and conduct of fathering: power shifts and awkward spaces

To say that the spaces of fathering are hidden or, at the very least, rendered in awkward ways, is not to say that there is significant debate about the context of fathering in contemporary society. Rather than emanating from discussions on masculine identity and politics, the last two decades' interest in fathering is fueled primarily by increased public employment for women and by unprecedented demographic changes in Western families. In a review of historical trends in notions of fatherhood, Pleck and Pleck (1997) argue that the father has evolved from the distant breadwinner of the nineteenth century, through the genial dad and sex role model of most of last century to today's father as equal co-parent. Geographer Linda McDowell (2000) points out that that these changes have triggered the emergence of new social cleavages between employed and unemployed men and women, as well as between men and women employed in different economic sectors and those working in the home. Paid employment and the assumed role of men as breadwinners and guardians of their families as a traditional key denominator for male identity is increasingly challenged. Although he does not disagree with this shift, Pierre Bourdieu (2001), in what was the last of his works to appear in English before he died, argues that the ascendancy of women in the productive sphere did not necessarily herald a redress of power imbalances. What is changed is a differentiation of new fields and subfields that reproduce the older structures: capitalism and patriarchy with new faces. When women get more education it is mainly in the humanities and not the sciences, notes Bourdieu; and when they enter management it is in its cultural and supportive aspects rather than finances and power. He posits a homology among fields so that the principles of sexual domination are repeated everywhere, but it may be argued that Bourdieu does not move far enough beyond structural principles (such as a binary parallelism between men and women) and his mechanisms of reproduction (such as an internalized habitus) so that he is powerless to explain the places where change takes place. Bourdieu's work is noteworthy in its specification of masculine domination, and the enduring sphere of patriarchy, but what needs some consideration are the ways that the spaces of fathering have changed in order to understand more fully if there are transformations in the idea of fatherhood. What questions of family politics and authority arise from these changes? What are the changing roles and relations of fathers in the lives of their children? How do fathers perform their identities in relation to children? How are the uneasy fits and awkward spaces of fathering manifest in everyday living?

These are heady questions that cannot be addressed without some discussion of the literature on male identities and the crisis of masculinities. In what follows, I draw on that literature to suggest that a large part of the above questions may be addressed with an explicit consideration of the ways fathers as well as mothers are impacted by what I call the 'patriarchal bargain'.

Victor Seidler (1997) argues that the 1990s saw a general withdrawal from the media conception of the 'nurturing, new man' that was in many ways driven by the discovery of a new market. This is a market niche that is elaborated upon geographically by Peter Jackson (1991; Jackson *et al.* 1999). In actuality, more often that not, middle-class men 'expressed an intention to be more involved with their children as fathers [but] found they did not have the time or the energy left after work' (Seidler 1997: 8). Arlie Hochschild (1997) and Susan Faludi (1999) note that for many men, life in the family is now harder work than life in paid employment. Additional evidence from some of my work with new parents in San Diego suggests that propelled by economic necessities, many fathers embrace the emotions of parenting without taking responsibility for the domestic labour that such emotions entail. A number of fathers in the San Diego family fantasies project committed more time to domestic activities (including being with the infant) than they did before the baby was born. That said, the complex changes accompanying the birth of a new child conflated overall responsibility and commitment with merely helping out at a day-to-day level. Awkward spaces and uneasy fits were created as men grappled with critical contradictions between the rational and egalitarian basis of shared responsibilities on the one hand, and the irrationality and emotion of the day-to-day work of child-care on the other (Aitken 1998: 87–9).

As the role of women has changed, leaving new spaces for male identities, McDowell (2000) argues that younger generations of men and boys do not easily fit into 'traditional' categories. At the same time, the creation of new social norms and values has not automatically led to an abandonment of the old, thus representing significant fields of conflict for the development of fathering identities. The family fantasies study took place at a time of economic recession in California. As part of an implicit patriarchal bargain, many fathers in the study torn between domestic responsibilities and the need to work longer hours to support their new child, chose paid employment. The bargain, put simply, is that patriarchal power relations are covertly embraced as cultural norms, and these norms are translated into fathers prioritizing their perceived role as principle breadwinner. At the level of everyday family life, contradictions between collective responsibilities and individual commitments are often systematized along traditional gender lines, even amongst men who actively engage in child-care and domestic activities. Thus, fathers are responsible for long-term career and financial support while mothers are responsible for long-term child-care. When men's work turns to child-care and women's work turns to financial support, many participants still felt that they were merely 'helping out'. A related identity problem is that fathering as a practice and fatherhood as an institution are difficult to define without recourse to mothering and motherhood. The men I interviewed in the family fantasies project who were solely responsible for child-care called themselves 'Mr. Mom' or 'house-husbands', reflecting and refracting their identities from the norms of mother and housewife (Aitken 2000). Some of these men joined support groups and networks of other full-time care-givers, but few felt comfortable in these female-dominated contexts.

The patriarchal bargain seems to prevail also in male-oriented support groups. In a study of a Lone Fathers Association in Newcastle, Australia, Hilary Winchester (1999a, 1999b) notes a reification of hegemonic forms of masculinity amongst members to the extent that her interviewees' views were 'profoundly misogynistic'. The most significant issues for these fathers concerned access to, and custody arrangements and maintenance payments for, their children. Most felt that the courts and their ex-wives treated them unfairly. Winchester's interviews raised important questions about inequitable access to legal aid and the vulnerability of all separated fathers to unsubstantiated allegations of child physical and sexual abuse (1999a). Such contexts, real or alleged, are an important part of the hidden spaces of fathering. Winchester points out further that the fathers in her study were frustrated by legal, institutional and familial structures predicated upon some very traditional ideas of fathers as the breadwinner of the family. She notes that 'while the individual circumstances of marriage breakdown vary enormously, the great chasm between the dream family and the reality of daily life suggests a tension leading to rupture' (Winchester 1999b: 92). There is a need to extend Winchester's concerns with a focus not only on the spatial rupture of families, but also attempts at either reconciliation or the formation of new family forms.

Empirical focus needs to move away from Winchester's lone (and disgruntled) fathers so that different perspectives on fathering may be embraced, and I suspect that the tension between 'dream' and 'actual' family life will play a significant role in future geographic debates on families. A further point that Winchester's work elaborates is the importance of local parenting geographies. The lone fathers in

Winchester's study aggressively re-asserted an instrumental, rough, controlling masculinity that drew in part from the maritime, industrial ambience of the city of Newcastle. Geographers Doreen Massey (1995) and Lawrence Berg (1994), in considering regional constructions of masculinity, demonstrate how such identities are contingent on place and time. Massey's study of Cambridge-based scientists highlights a view of masculinity that emphasized reason and control, while Berg finds that New Zealand masculinities are defined in part by a frontier ethic of wildness and danger. Sarah Holloway (1998: 31), in considering mothering cultures in Sheffield, identifies place-specific 'moral geographies of mothering' by identifying localized discourses concerned with what is considered right and wrong in the raising of children. What is needed are studies that highlight moral geographies of fathering around a specific context of family space.

It seems that the context of family space is misplaced for fathers precisely because identity politics do not account for the space and work of fathering in relation to the norms that continue to define fatherhood. The Western media is replete with stories of caring, nurturing, domesticated dads who are fighting the strictures of the workplace to spend the same amount of time with their children as mothers (e.g. Sauer 1993; Pryor 1999). In addition, a burgeoning social policy literature is focusing on men's co-parenting skills (Lamb 1997; Brandth and Kvande 1998). For the most part, media coverage and social science research compares levels of fathers' family involvement with mothers' involvement because mothers are the benchmark for norms in fathering (cf. Doherty et al. 1998: 278). Herein lie my concerns for the ways that fathers are defined in opposition to, or less than, mothers. I am not suggesting that fathering can be defined in isolation from mothering, but I am concerned that the idea of the father is constituted in parallel or in opposition to the idea of the mother and, as such, does not account for the imprecise and hesitant day-to-day work of fathering as resistance and negotiation.

I want to suggest that despite a growing literature espousing the value of the domesticated father, a powerful patriarchal 'idea' continues to influence the work and spaces of fathering. This idea is part of the patriarchal bargain. Some time ago, Ralph LaRossa (1988) suggested a consequential gap between what he called the 'culture' of fatherhood and the 'conduct' of fatherhood and, more recently, he demonstrated how this culture and conduct changes as social and political conditions change (LaRossa 1997). I think this is an important distinction because conduct suggests the work of day-to-day fathering whereas the culture of fatherhood suggests larger ideological concerns that are monolithic and perhaps unrealistic. But here I part company with LaRossa. In some of his work, LaRossa (1992) aligns with a belief that a 'new father' image is emerging that will eventually reconstitute fatherhood as a more nurturing institution. My concern is that research on 'the new father' tends to be uncritical and positive in its analysis of the contemporary culture of fatherhood while missing what is implied by the conduct of fathering (cf. Pleck 1987), with all its emotional trappings.

Ironically, a monolithic concept of 'the new patently domesticated father' was originally justified and supported by a feminist ideology that advocated men's involvement in parenting (Drakish 1989). It is perhaps worthwhile emphasizing that Doherty and her colleagues (1998) argue that monolithic notions do not account for differences in the kinds of work of fathering, and that work may indeed be more sensitive than mothering to contextual forces and day-to-day constraints. The new, domesticated father-ideal sidesteps the point that insidious patriarchal hegemony forces may create more obstacles than bridges for fathers. The image of the new father also assumes homogeneity in fathering roles and situational contexts that deny the importance of complex forms of fathering practiced differently in different spatial contexts.

Fathers and family space: a concluding comment

If fathering is to lose its awkwardness, I would argue that geographic research should be concerned with spaces of domination and exclusion as they relate to the idea of fatherhood *and* the emotional work of fathering.

It seems to me that the emergence of the importance of interior and emotional geographies provides a set of related theoretical ideas from which to draw. Of late, geographers have used psychoanalytic theory and psychotherapeutic practice to help analyze and understand the society-self relationships that, I argue, are fundamental to understanding boundaries and social exclusion (e.g. Bondi 1999; Nast 2000). There is some criticism that a focus on interiority detracts from enduring (larger, exterior) political and economic problems (Hamnett 1999; Martin 2001). Of particular note, however, for what I am trying to do here, is Steve Pile's (1996) defense of psychoanalytic theory's ability to highlight the 'politics of the subject' and Larry Knopp's (1992, 1995, 1998) attempts to examine the role of masculinity within the spatial dynamics of capitalism and globalization. These are huge agendas that go well beyond the scope of this chapter, and yet they simultaneously point to larger issues that revolve around the social and spatial construction of fathers through the work of fathering and the myth of fatherhood. Pile's work on interiority and Knopp's work on the construction of globalized and capitalized male subjects open possible doors for understanding fathering as a spatial practice that balances on larger familial, community and political issues.

The importance of emotional and political identity to larger social struggles suggests relations between identity and a politics of space. What emerged through the twentieth century was a strict division of gender roles that was precarious from the start because it gave fathers a choice that is apparently not afforded mothers. This, too, is part of a patriarchal bargain that hurts mothers and fathers equally. The empirical literature is clear that fathers in contemporary Western society both spend more time with their children than ever before and feel free to be more affectionate with them, or they simply walk out of their families. Thus, although large-scale structural forces are widening the gap between children and men, forces at the individual level sometimes work in the direction of closing the gap. There are emotional geographies here that require further academic attention from both the perspective of global economic restructuring and changes in local community and familial space.

I began this chapter with Laqueur's modest labour theory of fathering. I have tried to elaborate some mythic ideals that help structure the gender and spatial relations of fathering, and we still need to understand more fully how these get in the way of the day-to-day work of fathering. It is clear that the institution of fatherhood hinges on an 'idea' that does not embrace the 'fact' of fathering as a daily emotional practice that is negotiated, contested and resisted differently in different spaces. What is needed now and what I do not supply here are the voices of fathers from the spaces of their work.

References

Aitken, S.C. (1998) *Family Fantasies and Community Space*, New Brunswick, NJ: Rutgers University Press.

Aitken, S.C. (1999) 'Putting parents in their place: child rearing rites and gender politics', in Teather, E.K. (ed.) *Geographies of Personal Discovery: Places, Bodies and Rites of Passage*, London: Routledge.

Aitken, S.C. (2000) 'Fathering and faltering: "sorry, but you don't have the necessary accoutrements"', *Environment and Planning A* 32(4): 581–97.

Anderson, K. and Smith, S.J. (2001) 'Editorial: emotional geographies', *Transactions of the Institute of British Geographers* 26(1): 7–10.

Bell, D. and Valentine, G. (eds) (1995) *Mapping Desire: Geographies of Sexualities*, London and New York: Routledge.

Biller, H.H. (1993) *Fathers and Families: Paternal Factors in Child Development*, Westport: CT: Auburn House.

Blakenhorn, D. (1995) *Fatherless America: Confronting Our Most Urgent Social Problems*, New York: Basic Books.

Berg, L.D. (1994) 'Masculinity, place and a binary discourse of "theory" and "empirical investigation" in human geography', *Gender, Place and Culture* 1(2): 245–60.

Bernardes, J. (1985) 'Family ideology: identification and exploration', *Sociological Review* 33: 275–97.

Bloch, R. (1978) 'American feminine ideals in transition: the rise of the moral mother, 1785–1815', *Feminist Studies* 2: 100–26.

Bondi, L. (1999) 'Stages on journeys: some remarks about human geography and psychotherapeutic practice', *The Professional Geographer* 51(1): 11–24.

Bourdieu, P. (2001) *Masculine Domination*, Stanford, Ca: Stanford University Press.

Brandth, B. and Kvande, E. (1998) 'Masculinity and child-care: the reconstruction of fathering', *The Sociological Review* 46(2): 293–313.

Doherty, W.J., Kouneski, E.F. and Erickson, M.F. (1998) 'Responsible fathering: an overview and conceptual framework', *Journal of Marriage and the Family* 60: 277–92.

Drakish, J. (1989) 'In search of the better parent: the social construction of fatherhood', *Canadian Journal of Women and the Law* 3(1): 64–87.

Duncan, N. (ed.) (1996) *Body Space: Destablizing Geographies of Gender and Sexuality*, London and New York: Routledge.

Fagnani, J. (1993) 'Life course and space, dual careers and residential mobility among upper middle-class families in the Ile-de-France region', in Katz, C. and Monk, J. (eds) *Full Circles: Geographies of Women Over the Life Course*, New York: Routledge.

Faludi, S. (1999) *Stiffed: The Betrayal of the American Man*, New York: William Morrow and Company.

Furstenberg, F.F. (1988) 'Good dads – bad dads: two faces of fatherhood', in Cherlin, A.J. (ed.) *The Changing American Family and Public Policy*, Washington, DC: The Urban Institute Press.

Gottlieb, B. (1993) *The Family in the Western World: From the Black Death to the Industrial Age*, New York: Oxford University Press.

Hamnett, C. (1999) 'The emperor's new theoretical clothes, or geography without origami', in Philo, C. and Miller, D. (eds) *Market Killing: What the Free Market Does and what Social Scientists Should Do About It*, Harlow: Longman.

Hanson, S. and Pratt, G. (1995) *Gender, Work, and Space*, New York: Routledge.

Hawkins, A.J. and Dollahite, D.C. (eds) (1997) *Generative Fathering: Beyond Deficit Perspectives*, Thousand Oaks, Ca: Sage Publications.

Hawkins, A.J. and Palkovitz, R. (1999) 'Beyond ticks and clicks: the need for more diverse and broader conceptualizations and measures of father involvement', *The Journal of Men's Studies* 8(1): 11–32.

Hayford, A.M. (1974) 'The geography of women: an historical introduction', *Antipode* 6: 1–19.

Hochschild, A. (1997) *The Time Bind: When Work Becomes Home and Home Becomes Work*, New York: Metropolitan Books.

Holloway, S.L. (1998) 'Local childcare cultures: moral geographies of mothering and the social organization of pre-school children', *Gender, Place and Culture* 5(1): 29–53.

Hume, D. (1955) 'A treatise of human nature', reprinted from the original edition (1888) in three volumes, in Selby-Bigge, L.A. (ed.) Oxford: Clarendon Press.

Jackson, P. (1991) 'The cultural politics of masculinity: towards a social geography', *Transactions of the Institute of British Geographers* 24: 95–108.

Jackson, P., Stevenson, N. and Brooks, K. (1999) 'Making sense of men's lifestyle magazines', *Environment and Planning D: Society and Space* 17: 353–68.

Knopp, L. (1992) 'Sexuality and the spatial dynamics of capitalism', *Environment and Planning D: Society and Space* 10: 651–69.

Knopp, L. (1995) 'Sexuality and urban space: a framework for analysis', in Bell, D. and Valentine, G. (eds) *Mapping Desire: Geographies of Sexualities*, London: Routledge.

Knopp, L. (1998) 'Sexuality and urban space: gay male identity politics in the United States, the United Kingdom, and Australia', in Fincher, R. and Jacobs, J.M. (eds) *Cities of Difference*, New York and London: Guildford Press.

Kobayashi, A. (1994) 'Coloring the field: gender, "race", and the politics of fieldwork', *The Professional Geographer* 46(1): 73–80.

Lamb, M. (ed.) (1997) *The Role of the Father in Child Development* (3rd edn), New York: John Wiley and Sons, Inc.

Laqueur, T.W. (1990) *Making Sex: Body and Gender from the Greeks to Freud*, Cambridge, MA: Harvard University Press.

Laqueur, T.W. (1992) 'The facts of fatherhood', in Thorne, B. and Yalom, M. (eds) *Rethinking the Family: Some Feminist Questions*, Boston: Northeastern University Press.

LaRossa, R. (1988) 'Fatherhood and social change', *Family Relations* 36: 451–8.

LaRossa, R. (1992) 'Fatherhood and social change', in Kimmel, M. and Messner, M. (eds) *Men's Lives*, New York: Macmillan.

LaRossa, R. (1997) *The Modernization of Fatherhood: A Social and Political History*, Chicago, Illinois: University of Chicago Press.

Lewis, C., Feiring, C. and Israelashvilii, R. (1981) 'The father as a member of the child's social network', in Lamb, M.E. (ed.) *The Role of the Father in Child Development*, New York: John Wiley.

Longhurst, R. (2000) 'Geography and gender: masculinities, male identity and men', *Progress in Human Geography* 24(3): 439–44.

Los Angeles Times (2001) 'Supreme court justice rules on child custody and citizenship', 12 June, A9.

McDowell, L. (2000) 'The trouble with men? Young people, gender transformations and the crisis of masculinity', *International Journal of Urban and Regional Research* 24: 201–9.

Marsh, M. (1990) *Suburban Lives*, New Brunswick, NJ: Rutgers University Press.

Martin, R. (2001) 'Geography and public policy: the case of the missing agenda', *Progress in Human Geography* 25(2): 189–210.

Massey, D. (1995) 'Masculinity, dualisms and high technology', *Transactions of the Institute of British Geographers* 20(4): 487–99.

Nast, H. (2000) 'Mapping the "unconscious": racism and the oedipal family', *Annals of the Association of American Geographers* 90(2): 215–55.

Palkovitz, R. (1997) 'Reconstructing "involvement": expanding conceptualizations of men's caring in contemporary families', in Hawkins, A.J. and Dollahite, D.C. (eds) *Generative Fathering: Beyond Deficit Perspectives*, Thousand Oaks, Ca: Sage Publications.

Pile, S. (1996). *The Body and the City: Pyschoanalysis, Space and Subjectivity*, London and New York: Routledge.

Pleck, E.H. and Pleck, J.H. (1997) 'Fatherhood ideals in the United States: historical dimensions', in Lamb, M. (ed.) *The Role of Father in Child Development* (3rd edn), New York: John Wiley.

Pleck, J.H. (1987) 'American fathering in historical perspective', in Kimmel, M.S. (ed.) *Changing Men: New Directions in Research on Men and Masculinity*, London: Sage Publications.

Pryor, C. (1999) 'Welcome back to the family', *The Australian*, 19 May, p. 14.

Sauer, M. (1993) 'Chore war: which partner is doing what? How willingly', *The Union-Tribune*, 25 September, sec. E, p. 1.

Seidler, V.J. (1997) *Man Enough: Embodying Masculinities*, Thousand Oaks, Ca: Sage Publications.

Silverstein, L.B. (1996) 'Fathering as a feminist issue', *Psychology of Women Quarterly* 20(1): 3–37.

Winchester, H.P.M. (1999a) 'Interviews and questionnaires as mixed methods in population geography: the case of the lone fathers in Newcastle, Australia', *The Professional Geographer* 51(1): 60–8.

Winchester, H.P.M. (1999b) 'Lone fathers and the scales of justice: renegotiating masculinity after divorce', *Journal of Interdisciplinary Gender Studies* 4(2): 81.

Further reading

Hertz, R. and Marshall, N.L. (1998) *Working Families: The Transformation of the American Home*, Berkeley: University of California Press. This book offers stories of how families manage and how children respond to the rigors of their parents' lives, as well as broad overviews developed from survey and census data.

Lupton, D. and Barclay, L. (1997) *Constructing Fatherhood: Discourses and Experiences*, London: Sage Publications. *Constructing Fatherhood* provides an interdisciplinary analysis of the social, cultural and symbolic meanings of fatherhood in contemporary western societies.

18

STAGES AND STREETS

Reading and (mis)reading female masculinities[1]

Kath Browne

Summary

This chapter seeks to contribute to the literature on geographies of masculinities by offering a spatialized understanding of female masculinities. It focuses on drag kings and women who are mistaken for men to argue for a place within geographies of masculinities for female masculinity. Transgressions of sexed categories have been under researched and theorized within geographies. Conversely within queer theory, spatialities and contexts have not been sufficiently explored. This chapter seeks to open the channels for potential dialogues between queer gender theory and geographies by exploring drag kings in the context of on-stage/off-stage performances and the everyday disruptions of women who are mistaken for men.

- Where girls will be boys: queering performance spaces
- Streetlife: women who are mistaken for men.

Introduction

Geographies of masculinities have yet to explore masculinities beyond 'men's' bodies. This chapter queers the supposed link between male bodies and masculinities by investigating masculinities and 'women's' bodies. However, I do not have the space to include all aspects of female masculinity that Halberstam (1998) details in her book. I will not explore 'butchness', male to female transexuality or male impersonation in any detail (along with Halberstam see Feinberg 1993; Lee 2001; Maltz 1998; Munt 1998; Middlebrook 1998 for more detailed investigations of these). Instead, masculinities, as diverse and diffuse, are used here specifically to understand how women consciously, overtly and hyperbolically perform 'maleness' (drag) and the male readings of women's bodies (women who are mistaken for men).

In this chapter I seek to open up dialogues between queer gender theorists and geographies. Gender disruptions are underexplored in geographic research with 'queer' usually understood implicitly as sexually marginalized or 'deviant' individuals (exceptions include Cream 1995; Namaste 1996). This chapter addresses this lacuna by exploring the readings of queer performances and queer lives arguing that the places where gender transgressions occur are important to their (mis)readings, in this case stages and streets (or everyday spaces). The chapter is divided into two sections. The first investigates the drag king scene focusing on the potential hegemonic readings of the on-stage/off-stage disjunctures of gender and the messy female masculinities that transgress on-stage/off-stage binaries. In contrast to these playful masculinities the second section explores upsetting (mis)readings of bodies where women are mistaken for men.

Where girls will be boys: queering performance spaces

> A drag king is a performer who makes masculinity into his or her act (yes there can be male Drag Kings).
>
> (Volcano and Halberstam 1999: 36)

Drag kings parody and hyperbolize masculinities in numerous ways and these can be related to different forms of masculinities (including 'black' masculinities, misogynistic masculinities, homophobic masculinities, gay masculinities) (Volcano and Halberstam 1999; Hausten 1999). Performance routines also take a variety of forms including lip-synching, dancing and enacting particular male personas on-stage. In Figure 18.1, for example, the Chicago drag kings enact their version of jailhouse rock, playing both with the persona of Elvis and imprisoned masculinities.

'Masculine' stylizations of women's bodies have often been termed 'butch' rather than being read as drag king performances.[2] In contrast to the drag king, other manifestations of female masculinities do not don masculinity as a performance or wear male clothing as drag, rather 'they embody masculinity' (Halberstam 1998: 241). At times clear lines have been drawn between drag kings and those that embody masculinity. For example, the separation of drag, male impersonation, passing and butch identities is clear in Volcano (DLV) and Halberstam's (JH) (1999) discussion with Mo B Dick (MF), a drag king and organizer of a drag king club in San Francisco:

DLV: Then what are the competitors who you think are not Drag Kings?

MF: I think they are very butch women . . . one night at HerShe bar I was so insulted because a very baby butch won, and I had gone to great lengths to prepare: I wore a killer shark skin suit, I did my pompadour, you know, I was stylin' . . . and I was working it and she just went up there on a whim in an outfit she wears every day . . .

JH: so one contest is about how butch they are and whether they pass every day; the other one is about how well they imitate maleness.

(Volcano and Halberstam 1999: 112)

Figure 18.1 The Chicago Drag Kings (*source:* R.J. Spencer).

Mo B Dick is clearly aggrieved by the lack of effort the 'baby butch' put into her 'act'. For Mo B Dick there is a clear distinction, articulated by Judith Halberstam, between those who do masculinity on-stage (both in terms of how one plays a man and how one dresses) and those who are 'masculine' in their everyday lives. Mo B Dick goes on to describe how he/she can also be feminine when outside of drag. This movement across and between gendered categories differs from those who enact butchness on the street as well as on stage. In this interpretation, those who do not make an effort to hyperbolize or parody maleness on stage are not 'drag kings' per se. There must be a performance and one feels from Mo B Dick's comments that an element of separation between on-stage and off-stage is also necessary.

Queer theorizations have argued that all gendered identities are acts of passing (Butler 1990, 1993; Ahmed 1999). Whilst queer suggests that we are what we do (Butler 1990), the literature on drag kings, and drag kings themselves can separate hyperbolic drag king performances from butch women, female to male transsexuals and other forms of masculine performativities (Hausten 1999; Halberstam 1998; Volcano and Halberstam 1999). This separation is spatially delimited. Where the masculinities are staged it is 'drag', because in these spaces the *performance* and parody of masculinities is centralized (Halberstam 1998). In contrast, enacting a butch female identity and/or passing as male in everyday spaces are something different. In this sense drag performances are somewhat distanced from the mundane daily performativities of gender and sex. Yet drag kings draw on these normative masculinities for the purposes of parody.

Drag disruptions and the limits of the stage

Having separated drag from other forms of female masculinities, it is important to explain the focus on drag. Drag is frequently seen as the activism behind queer theorizations of fluid sexes and these discussions have centralized drag queens' appropriations of femininities. However drag kings can also contest dichotomous sexes:

> The drag king performance, indeed, exposes the structure of dominant masculinity by making it theatrical and by rehearsing the repertoire of roles and types on which such masculinity depends.
>
> (Halberstam 1998: 239)

Drag kings and queens by enacting femininities and masculinities for the purposes of parody, reverence or humour, contest the 'realness' of gendered identities and may challenge the natural association of gender roles and types with specific bodies (see Butler 1990). Understanding all genders (and sexes) as imitations of an imaginary real, and as performativities that require continual reiteration, both drag kings and drag queens have the potential to expose the 'realness' of sex as a reproduction. Through parodying and hyperbolizing the performance of femininity and masculinity, the performative contingency of men and women is 'exposed'.

Butler (1993) critiqued interpretations of *Gender Trouble* that read 'drag' as *the* tool for disrupting sexed binaries. She argued that resistances such as drag can be recuperated within normative codes and 'there is no guarantee that exposing the naturalized status of heterosexuality will lead to its subversion' (Butler 1993: 231). The disjuncture that queer readings of drag draw upon relies on the performance being different from the 'real' and in this case drag kings playing with masculinity deliberately rather than 'doing' their sexed identities. This is obvious in the separation of passing, butch and other forms of female masculinities from drag king performances. Moving from the intention of the actor to the reading of the audience and taking account of the spatial, it can be argued that due to drag acts being on stage (or in performance spaces) the 'real' gender of the actor may remain unquestioned (by actor and/or audience). (S)he is just acting, hyperbolically or otherwise, as a member of the dichotomously opposite gender. Gender is 'played with', and the spatial as well as imaginative distance between 'the staged performance' and 'the real embodiment' may enable dominant readings of sex to be maintained. Consequently, contextual readings of these performances may distance the on-stage 'unreal' gender performance from off-stage 'real' sexed enactments.

Challenging gender transgressions(?)

I began this chapter with the intention of arguing that drag acts, whilst showing the performativities of gender on stage, may be read as 'unreal sex' because of *where* they are performed. I wanted to think through the spatial separations of reading and performing, intention and interpretation, gender 'play' and 'real sex'. I wondered how much of the gender transgression can be dismissed as entertainment, humour, surreal not 'real' because the 'where' of the performance influenced the reading of that performance. Furthermore, I could see that drag, by deliberately 'playing' with gender roles and distancing 'real' sex from 'performed' gender, enabled readers to recuperate hegemonic dichotomies. I would agree with Butler that drag performances are not necessarily subversive but they do offer transgressive possibilities and I would add that the spatial is important in exploring the readings of gender disjunctures. However, I was/am uneasy setting up a real/unreal dualism.

It became clear in reading for this chapter that within female masculinities messy spatial crossings can contest the real/unreal dualism. Volcano and Halberstam (1999) discuss the place based differentiations of drag king scenes comparing San Francisco, London and New York.[3] They argue that whilst in New York few women express transgender or butch identities off-stage, seeing drag as a theatrical performance, in London and San Francisco the on-stage/off-stage personas are not so clearly delimited. Thus, a distance does not necessarily exist between the 'real' of passing or of being butch and the drag king shows.

Due to the late development of drag king scenes[4] and the presence of diverse forms of female masculinities, particularly butch identities, the boundaries that form these scenes are permeable. As Halberstam contends: 'The blending of onstage drag and off-stage masculinity suggests the line between male drag and female masculinity in a drag king club is permeable and permanently blurred' (Halberstam 1998: 244).

Drag king shows can offer safe spaces in which to perform alternative gendered identities and male personas may not be confined to the stage. Some drag kings pass as men in every day life. It is the messiness, created through these spatial crossings, overlaps and intersections, that (re)creates sexed ambiguities challenging dichotomous sexes. Drag kings, butch women and female to male transsexuals can refuse to perform neat on-stage/off-stage identities (Volcano and Halberstam 1999: 32). In Figure 18.2, Sid Viscous sits comfortably in a bar drinking, transgressing the on-stage 'Sid' and off stage 'Jude'. Thus, there can be continuities between 'doing' gender transgressively and 'performing' drag, and these continuities, along with distances between 'doing' and 'acting', contest the safe assumption that, whilst 'girls will be boys' they will always return to (hetero)sexually attractive (and available) women (Halberstam 1998).

Blurring the boundaries of man/woman offstage may contest the man/woman dichotomy through which normative everyday spaces are (re)produced as well as empowering individuals to move beyond proscribed sexed boundaries on or off stage. In other words taking account of the spatial illustrates the complexity of reading sexed embodiments and identities, further contesting the dichotomization of man/woman (see also Bell et al. 1994).

Figure 18.2 Sid Viscous (*source:* Fionnuala Murphy).

Streetlife: women who are mistaken for men

The chapter will now move away from the fantastical stages on which drag is often performed or the clubs where drag shows take place, to the mundane everyday spaces where women are mistaken for men. In these contexts readings of bodies are problematic as individuals do not fit within the presumed natural dichotomy of man/woman. Sights of sexed bodies are important as they can play a part in (re)forming the identities and embodiments of those who are viewed (Bornstein 1995; Straayer 1997). These readings can thus move individuals between the categories of man and woman.

This section of the chapter is based on my doctoral research conducted with 28 non-heterosexual women. The research was designed to examine non-heterosexual women's mundane materialities. Being mistaken for a man was an emergent theme in the study. Nine women described experiences of being mistaken for a man in six focus groups, two coupled interviews and eight individual interviews. These stories were often emotive but the experiences went unnamed and often unacknowledged (see Browne 2003, 2005). Stevi illustrates the importance of being read as the sex one lives and the dichotomization of these readings:

Me and Susan [ex-girlfriend] just went away for a week and just went into this little tiny just a little café. . . . at the end [of the meal] she [the waitress] came over with the bill. Actually she waited until we gave her the money and she went 'thank you very much ladies, sir, oh whatever you are'. 'Thank you very much, good night'. Now I that's just disgusting, I think. That, you know that upsets me, you know 'whatever you are'.

(Stevi and Virginia, coupled interview)

Ward (1765) asserted that 'he must represent a male; she a female; and it, an object of no sex' (quoted in Bodine 1998: 129). Stevi and her partner did not fit into the waitress's dichotomous groupings, 'ladies' or 'sir' and so they became dehumanized as 'whatever'. This 'what' rather than 'who' indicates an object instead of a person. Without a sex individuals can no longer be human. Consequently Stevi and her ex-partner are not intelligible as human outside of easily readable gender categories. The experiences of being mistaken for a man move the individuals involved across and between the man/woman dichotomy. This is not simply a theoretical debate; it is often an emotionally charged experience, one that Stevi describes as 'upsetting' and 'disgusting'. Everyday life can problematize the 'playful' image of gender transgressions that is often presented within queer theory and associated with drag performances (see, for example, Queen and Schmeil 1997). Furthermore, gender disjunctures are not always 'safe' nor do they only occur in the protected environments of lesbian/gay clubs or on drag king stages.

The unintelligibility and contestation of what is often perceived as 'natural' permanent and fixed, i.e. sex, resulted in participants in my doctoral research looking to reaffirm their identities as female. Bodies were used to (re)place individuals within the category 'woman':

KB: Do you get mistaken for a lad?

Julie: Ehm I haven't recently. When I used to wear my cap I did. But I don't now.

KB: Oh that's cool.

Julie: Do I look like a bloke?

KB: No.

Julie: That's alright.

KB: But no you said in the focus group that you do sometimes get mistaken for . . .

Julie: I have, I have in the past been mistaken for . . .

KB: How does it make you feel?

Julie: [makes noise] Don't know mate. There's a couple of breasts there. Just trying to make a big point I've got breasts and I'll stick them out and shove em in their throat, damn it, shove it down their throats so then they'll know.

(Julie, individual interview)

Julie illustrates the sensitivity of reading female masculinity. When I ask if she has been mistaken for a 'lad' (a term used in the focus group with Andie), she asks if she looks like a 'bloke'. Not believing that she does, and not willing to upset her I immediately say no (in addition the use of the phrase 'cool' implies that it is a good thing she is no longer mistaken for a man). My reaction illustrates that it is an insult to be read and categorized outside your lived gender. Her relief at being read in her correct gender is illustrated through the phrase 'that's alright'. This contests Devor's (1987) assertion that within a patriarchal society 'there would be less offence in erring in the direction of affording someone higher status [male] rather than the lower [female]' (Devor 1987: 20). The interactions between Julie and myself, illustrate that the offence can lie in the disjuncture between self-perceptions and gender ascriptions by others (see Halberstam 1998). This may, however, not be the case in gay male communities where the term 'she' and other inverted appellations are at times acceptable (see Bunzel 2000). Perhaps the insult of being called a 'man' pertains in part to female masculinities being labelled 'butch' rather than embodying male terminology; in contrast effeminate and gay men more readily appropriate feminine pronouns. The result is perhaps a more jarring insult, a questioning of one's appearance ('do I look like a bloke?') followed by a (re)inscription of sexed identity within 'woman' ('There's a couple of breasts here').

Julie argues emotively that her breasts mark her as female. When she is mistaken for a man she will use her body to prove that she is a woman thereby demonstrating the salience of the body and the importance of being read as a woman, as well as 'doing' femininity. Andie verbally confronts the questioning of her sex:

> KB: Have you ever had any negative experiences or anything like that in relation to food?
>
> Andie: Not really apart from the amount of times that I have heard 'and you sir?' I am going (looks over her shoulder) looking behind me going, 'is there a bloke here? Are you a bloke? I don't think so'. Its like 'alright M [friend] are you a bloke?' 'NO. I don't think I am. No, I'm not' [laughter].
>
> (Andie and Julie, focus group)

When asked about her negative experiences in a restaurant Andie immediately refers to when she is called 'sir' and discursively constituted as male. Clearly Andie finds problematic the reading of her as male, perhaps because it challenges her self-identity as female. To the question 'are you a bloke?' she answers 'NO. I don't think I am. No, I'm not'. Her reaction illustrates the necessity of denying the other, in this case male, in order for her embodiment and identity to make sense within the dichotomous terms of sex. By discursively (re)placing herself as female she becomes intelligible within liveable society (see Butler 1997). The laughter in the focus group demonstrates the assumed impossibility of moving between the categories of man and woman. The reader has simply got it wrong and needs to be corrected. The emphatic affirmation of Andie's sex within the category woman is indicative of the importance she places on sexed categories and her position within this dichotomy. Pat uses similar tactics to challenge inquiries about her sex:

Pat: If people are like 'are you a man?' and I'm like 'I don't know, are you?' and just like take the piss out of them. . . . [laughter] That's the best way to get over, if you are embarrassed yourself is just to embarrass other people.

KB: . . . how does it make you feel when you are mistaken for a bloke?

Pat: Well obviously not very nice, yeah it is very embarrassing.

(Pat, individual interview)

Pat contests the problematization of her sex. She replies 'I don't know', challenging the notion of a known sex that one should actively defend. Subsequently, she turns the question on the person who asked it contesting the secure assumptions this person holds of their sexed identity. If this person is 'obviously' female, Pat questions their place in this category, if male, Pat is challenging their masculinity. Either way it is discomforting for both parties and Pat is no longer alone in her embarrassment. Interestingly, Pat illustrates that whilst transgressing normative gender can be uncomfortable, the process of contesting these readings can be personally empowering. Consequently, the interplay between reader and actor in producing sex can be negotiated rather than imposing strict categories. It should be stressed that these dialectics are 'not very nice' and can be 'very embarrassing' for those accused of crossing the boundaries of man/woman (see Browne, forthcoming a).

Women who are mistaken for men can contest the supposed 'natural' links between sex and how one's body is read. These moments of transgression, where women's bodies are read as male, may challenge the illusion of fixed sexes. Being mistaken for a man differs from intersexuality, transsexualism, transgenderism as well as drag. It is a (mis)reading rather than an intentional challenge of gender norms and boundaries, moreover the fluidity of sexed readings is also important. At different times in different places these individuals may be read as either a man or a woman. This can be understood as unintentional 'resistance'. However, importantly, this dissonance is recuperated not by those external to body sites, as may be the case in drag (Butler 1990, 1993), but by those who are 'mistaken'. The frequent recourse to breasts (regardless of size) in this study by those who are (mis)read reconstitutes the sites of bodies within intelligible discourses. In this way the 'mistake' is 'corrected' but, through the process of disjuncture, the assumption of a binary gender system, where opposites 'naturally' occur, is contested. It is important to reiterate that although gender disruptions can be read as rendering man/woman fluid and performative, the (re)formation of these categories can be painful, disconcerting and discriminatory for those whose embodiments do not 'fit' dichotomous gender norms (see Browne, 2004).

Conclusion

Geographies of masculinities, along with wider geographies of gender, often do not question the assumptions of masculine-man, feminine-woman. This chapter has sought to disrupt these neat linkages, contesting the naturalized association of men with masculinity. Through an exploration of two forms of female masculinity it has been argued that spaces where (mis)readings occur are important to understanding gender dislocations, instabilities, fluidities as well as discriminations. Taking the spatial into account, readings of drag performances on stage can be distanced from 'real' gender performativities off-stage. This may in part be due to interpretations of where these 'unnatural' acts take place, specifically in 'unreal', theatrical and dramatic spaces. When moving to more 'banal' spaces, the chapter explored readings that are themselves transgressive in that they cross the sexed boundaries of man/woman and 'mistake' women for men. In everyday sites the slippages between gender and sexed categories cannot be distanced from everyday life. However, 'real' body sites (vaginas, breasts, penises and so forth) are employed by people themselves, as well as those reading their bodies, to recuperate (sometimes violently and abusively) dichotomous divisions between men and women.

Further explorations of spatialized readings could lead to important insights into the queer disruptions and failures of gender iterations as well as the hegemonic policing that keeps dichotomous genders and sexes 'in place'. Taking contexts seriously requires an exploration of the (re)creation of bodies and identities beyond the individual and between people (see Rose 1999). Perhaps the disruptions to sexed categories are continually and un/intentionally located in the spaces between us as well as in the settings that these interactions (re)make. Geographies of masculinities are well placed to explore these disjunctures alongside the hegemonic practices that maintain sexed dualisms.

Acknowledgements

I would like to thank all the participants who took part in my doctoral research and particularly the nine women whose stories form the empirical aspect of this chapter. My sincere gratitude goes to Cara Aitchison who was a constant source of advice and support throughout the doctoral process. I would like to thank Sid Viscous and Spencer for the use of the photographs. Thanks also go to Kathrin Hörschelmann and Bettina van Hoven for inviting me to contribute to this book and their helpful comments.

Notes

1 An earlier version of this chapter was presented as a paper at the RGS/IBG Annual Conference 2003, London, in the 'Wear in the Word: (Re)materializing geographies of gender and sexualities' session.
2 Conversely drag queens have long historical connections to gay male cultures and increasingly permeate more mainstream stages such as television and heterosexualized venues of

restaurants, bars and clubs. There is a long tradition of drag queens and 'men' in 'women's' dress are often likely to be termed drag queens (Volcano and Halberstam 1999; Halberstam 1998). Drag queens have been defined as persons who perform femininity as an overt and conscious act and impersonate women during these performances (Volcano and Halberstam 1999).

3 Halberstam (1998) also differentiates drag king contests, which attracted more 'butch' women and drag king clubs where performers recognize and understand themselves as drag kings.

4 The development of the drag king scene has not mirrored that of the drag queens such that drag kings do not have the same history, meaning and value (Willcox 2002). The drag king scene, in contrast to male impersonation, butch identities and transgenderism, is seen as a recent manifestation of the late twentieth century (see note 2 above).

References

Ahmed, S. (1999) 'She'll wake up one of these days and find she's turned into a nigger: passing through hybridity', *Theory, Culture and Society* 16(1): 87–106.

Bell, D., Binnie, J., Cream, J. and Valentine, G. (1994) 'All hyped up and no place to go', *Gender, Place and Culture* 1(1): 31–47.

Bodine, A. (1998) 'Androcentricism in prescriptive grammar: singular "they" sex-indefinite "he", and "he or she"', in Cameron, D. (ed.) *The Feminist Critique of Language: A Reader*, London: Routledge.

Bornstein, K. (1995) *Gender Outlaw: On Men, Women and the Rest of Us*, New York: Routledge.

Browne, K. (2003) 'Negotiations and fieldworkings: friendship and feminist research', *ACME: An International E-Journal for Critical Geographers*, 2(2): 132–46.

Browne, K. (2004) 'Genderism and the bathroom problem: rematerializing sexed sites, recreating sexed bodies', *Gender, Place and Culture* 1193): 331–46.

Browne, K. (2005) 'Snowball sampling: using social networks to research non-heterosexual women', *International Journal of Social Research Methodology*, forthcoming.

Bunzel, M. (2000) 'Inverted appellation and discursive gender insubordination: an Austrian case study in gay male conversation', *Discourse and Society* 11(2): 207–36.

Butler, J. (1990) *Gender Trouble*, London: Routledge.

Butler, J. (1993) *Bodies that Matter*, London: Routledge.

Butler, J. (1997) *The Psychic Life of Power: Theories in Subjection*, Stanford: Stanford University Press.

Cream, J. (1995) 'Re-solving riddles: the sexed body', in Valentine, G. and Bell, D. (eds) *Mapping Desire: Geographies of Sexualities*, London and New York: Routledge.

Devor, H. (1987) 'Gender blending females: women and sometimes men', *American Behavorial Scientist* 31(1): 12–40.

Feinberg, L. (1993) *Stone Butch Blues*, Ithaca, NY: Firebrand.

Halberstam, J. (1998) *Female Masculinity*, London: Duke University Press.

Hausten, L. (1999) *Gender Pretenders: A Drag King Ethnography*, Masters: Columbia University in the City of New York.

Lee, T. (2001) 'Trans(re)lations: lesbian and female to male transsexual accounts of identity', *Women's Studies International Forum* 24(3–4): 347–57.

Maltz, R. (1998) 'Real butch: the performance/performativity of male impersonation, drag kings, passing as male, and stone butch realness', *Journal of Gender Studies* 7(3): 273–86.

Middlebrook, D.W. (1998) *Suits Me: The Double Life of Billy Tipton*, New York: Mariner Books.

Munt, S.R. (1998) *Butch / Femme: Inside Lesbian Gender*, London: Cassell.

Namaste, K. (1996) 'Genderbashing: sexuality, gender, and the regulation of public spaces', *Environment and Planning D: Society and Space* 14(2): 221–40.

Nelson, L. (1999) 'Bodies (and spaces) do matter: the limits of performativity', *Gender, Place,* Queen, C. and Schmeil, L. (1997) *Pomosexuals: Challenging Assumptions about Gender and Sexuality*, San Francisco: Cleis Press.

Rose, G. (1999) 'Performing space', in Massey, D., Allen, J. and Sarre, P. (eds) *Human Geography Today*, Cambridge: Polity Press.

Straayer, C. (1997) 'Transgender mirrors: queering sexual difference', in Holmlund, C. and Fuchs, C. (eds) *Between the Sheets, in the Streets: Queer, Lesbian, Gay Documentary*, London: University of Minnesota Press.

Volcano, D.L.G. and Halberstam, J. (1999) *The Drag King Book*, London: Serpent's Tail.

Willcox, A. (2002) 'Whose drag is it anyway? Drag kings and monarchy in the UK', *Journal of Homosexuality* 43(3–4): 263–84.

Further reading

Cream, J. (1995) 'Re-solving riddles: the sexed body', in Valentine, G. and Bell, D. (eds) *Mapping Desire: Geographies of Sexualities*, London and New York: Routledge. This chapter introduces a range of gender transgressions and questions the continuities between sex and gender.

Halberstam, J. (1998) *Female Masculinity*, London: Duke University Press. Halberstam's text is a key text that explores many different facets of female masculinity. It is the core text in this area of study.

Namaste, K. (1996) 'Genderbashing: sexuality, gender, and the regulation of public spaces', *Environment and Planning D: Society and Space* 14(2): 221–40. Namaste explores the transgression of normative relations of gender and sex and how public space is (re)defined through gendered enactments.

Volcano, D.L.G. and Halberstam, J. (1999) *The Drag King Book*, London: Serpent's Tail. This book offers theory, interviews and pictorial representations of drag kings. It documents 'key' drag king lives and performances combining analysis and description of drag kings and drag.

INDEX

Printed and bound by CPI Group (UK) Ltd, Croydon, CR0 4YY

01/11/2024

01782635-0007